CONCRETE DAM INSTRUMENTATION MANUAL

by
Charles L. Bartholomew
and
Michael L. Haverland

October 1987

United States Department of the Interior ★ Bureau of Reclamation

ACKNOWLEDGMENTS

This manual was prepared under the general direction of Dewayne L. Misterek, Chief of the Structural Behavior Branch, Division of Dam Safety, Bureau of Reclamation, Denver, Colorado. Dr. Charles L. Bartholomew, the principal author, was employed by the Structural Behavior Branch for the summers of 1985-87; he is a professor, and Chairman of the Department of Civil Engineering at Widener University in Chester, Pennsylvania. Michael L. Haverland, the contributing author, is the Head, Concrete Dam Instrumentation Section at the Bureau of Reclamation. The review comments of D. L. Misterek and J. L. Kinney, Structural Behavior Branch, are greatly appreciated by the authors.

Certain sections of this manual were modified from the "Embankment Dam Instrumentation Manual" published by the Bureau of Reclamation in January 1987. In addition to Dr. Bartholomew, Bruce C. Murray, Head, Embankment Dam Instrumentation Section, and Dan L. Goins, also of the Embankment Dam Instrumentation Section, were authors of that publication.

Portions of chapter 6 were written by Andy Viksne, Head, Geophysics Section; and portions of chapter 7 were prepared by T. E. Backstrom, Head Chemistry Section; and by Lynn Carpenter and Jay N. Stateler of the Structural Behavior Branch. Significant contributions were made by James Kinney, Michelle Barry, Dennis Cox, Margaret Crawford, Martha Jong, and John Soderquist of the Concrete Dam Instrumentation Section in the preparation of the majority of the figures.

PREFACE

The Concrete Dam Instrumentation Manual has been prepared to make available to designers, engineers, instrument readers, dam operators, and dam safety personnel the information needed for installation, operation, and analysis of instrumentation systems. The manual is primarily intended for Bureau of Reclamation personnel. Other agencies or individuals, either in the United States or in foreign countries, who are engaged in design or construction of concrete dams or in the safety evaluation of dams may also find the information to be useful.

Material used in preparing the manual has been drawn from many sources and an attempt to reference these sources has been made where appropriate.

CONTENTS

CONTENTS—Continued

CONTENTS—Continued

CONTENTS—Continued

CONTENTS—Continued

CONTENTS—Continued

CHAPTER 6—VIBRATION MEASUREMENT DEVICES

A. HISTORY AND DEVELOPMENT OF DEVICES

B. CURRENTLY USED DEVICES

CONTENTS—Continued

TABLES

CONTENTS—Continued

FIGURES

CONTENTS—Continued

CONTENTS—Continued

Chapter 1

INTRODUCTION TO INSTRUMENTATION

A. General Considerations

1.1 Purpose.—Dams are designed and constructed to impound storage reservoirs; therefore, dams are usually the key structures of projects designed for development of river basin potential for irrigation, water supply, hydroelectric power generation, flood control, recreation, navigation, and other significant economic benefits. Dams are expected to safely withstand, over many years, the potentially enormous forces created by the impounded water.

The sudden release of impounded water in the event of a dam failure has tremendous potential for destruction of life and property located downstream. Therefore, the proper and safe functioning of a dam is an extremely important matter of economic benefits and public safety. Considerations of the destructive effects of a dam failure in terms of loss of life, damage to property, concern for public welfare, and negation of planned benefits of the project make it imperative that means are available for gathering information to assess the performance and continuing assurance of the safety of the dam during construction, first filling of the reservoir, and during long-term service operation. The most important objective of an instrumentation monitoring program for a dam is to provide a major portion of the necessary structural behavior information to compliment an adequate dam safety inspection and surveillance program.

1.2 Need.—Factors or quantities that should be monitored in concrete dams include structural displacements, deformations, settlement, seepages, piezometric levels in dam foundation, and uplift pressures within the structure foundation. Total movements are monitored as well as relative movements between portions of a dam. A wide variety of instruments are utilized in a comprehensive monitoring program. The full range of instrumentation used by the Bureau is discussed in this manual. The principal objectives for using instrumentation may be generally grouped into diagnostic, predictive, legal, and research categories.

    a. *Diagnostic.*—(1) *Verification of Design Parameters.*—Instrumentation often plays a major role in verifying design parameters with observations during construction, which frequently enables the engineers to determine the suitability of the design as the construction progresses. Furthermore, the information also aids in modifying purely theoretical treatments by incorporating the effects of actual field conditions. Design of dams generally entails a rigorous and sometimes complex study of forces based on rather conservative assumptions concerning material characteristics and structural behavior. These assumptions are made to provide for unknowns or uncertainties in the design.

Observations from the monitoring systems and an assessment of the influence of the various factors on structural performance of the dam can help mitigate these unknowns, thereby leading to progressive refinements and improvements in analysis techniques and future designs.

    (2) *Verification of Suitability of New Construction Techiques.*—Experience has shown that most new or modified construction techniques are not readily accepted by the construction or engineering professions until proven satisfactory on the basis of actual performance. Data obtained from instrumentation can aid in evaluating the suitability of new or modified techniques.

    (3) *Diagnosing Specific Cause of an Adverse Event.*—When a failure, partial failure, or severe distress has occurred at a damsite, data from an instrumentation system can be extremely valuable in the determination of the specific cause or causes of the event. Also, instrumentation is often installed just prior to or during remedial work at a site to determine the effectiveness of the improvements and the effect of the improvements on the existing dam conditions.

    (4) *Verification of Continued Satisfactory Performance.*—An instrumentation system that consistently yields data that indicates that the dam is performing in a satisfactory manner may initially appear to be unnecessary. However, such information can prove to be valuable should some future variation in data occur signaling a possible problem, and because such satisfactory performance is, in itself, valuable for future design efforts. Satisfactory performance is also very reassuring to the owner or operator of a dam.

b. *Predictive*.—It is important to be able to utilize instrumentation data during accumulation so that informed, valid predictions of future behavior of the dam can be made. Such predictions may vary from satisfactory performance to an indication of severe future distress which may become threatening to life or safety and necessitate remedial action.

c. *Legal*.—Valid instrumentation data can be valuable for several reasons ranging from simple determination of actual construction quantities for construction pay estimates to the establishment of an information data bank for possible later use in litigation. Damage claims arising from dam construction or from adverse events can reach millions of dollars. Instrumentation data can also be an aid in determining causes of adverse events so that proper legal adjudication can be accomplished.

d. *Research*.—To understand the complex nature of the multitide of forces acting in an usually interdependent manner on a dam, it is desirable to study the performance of existing dams and the instrumentation data generated thereon, which should provide quantifiable information for use on future designs. Such research has led to advances in construction techniques, improved and innovative design concepts, and a better understanding of failure mechanisms.

1.3 Concrete Dam Failures.—Jansen [1][1] presented a number of studies listing and discussing dam failures that have occurred over about the last 900 years. Of more than 150,000 dams that represented hazards to life or property, only about 2,000 have failed over the past 9 centuries, and most of these failures were not major dams. During the 20th century, there have been about 200 notable failures resulting in the loss of over 8,000 lives. Of these 200 failures, less than 40 have been concrete or masonry dams. Jansen lists detailed information on 15 significant concrete and masonry dam failures, which are summarized in table 1-1. The majority of these failed facilities have been successfully rebuilt and are operating in an adequate manner. This table shows that (eliminating acts of war) nearly all of these failures resulted either directly or indirectly from foundation or abutment problems. Therefore, a number of monitoring instruments have been devised to detect seepage, hydrostatic pressures, deformations, and movements of foundations and abutments.

Common evidences of deficiencies in concrete dams include:

- *Stress and Strain.*—Cracks, crushing, or offsets in concrete monoliths, buttresses, face slabs, arch barrels, galleries, operating chambers, and conduits; stress and temperature cracking patterns in buttresses, pilasters, diaphragms, and arch barrels; or stress decline in posttensioned anchorages and tendons.
- *Instability.*—Excessive or unevenly distributed uplift pressures; differential movement of adjacent monoliths, buttresses, arch barrels, or face slabs; movement along construction joints; or uplift on horizontal lift surfaces revealed by seepage on the downstream face or in galleries.
- *Seepage at Discontinuities and Junctures.*—Embankment wraparound sections, waterstops in monoliths and face slabs, or reservoir impounding backfill at spillway control sections and retaining walls.
- *Foundations and Abutments.*—Piping of material from solution channels or rock joints; clogged drains; movement at faults or shear zones; sliding along bedding planes; or consolidation of weak strata due to the mass of the dam and reservoir, or in a horizontal plane due to thrust from an arch dam.

1.4 Instrumentation Philosophy.—a. *General*.—As previously discussed, there are many reasons for installing instrumentation in both new and existing dams. The question on the number, type, and location of instruments required at a dam can only be addressed effectively by the fortuitous combination of experience, common sense, and intuition. Dams represent unique situations and require individual solutions to their instrumentation requirements. The instrumentation system design; therefore, needs to be conceived with care and consideration for the site-specific geotechnical conditions present in the foundation and abutments of the dam. In general, it has been found that an adequate, but cost-effective, instrumentation program at a new dam will constitute about 1 percent of the total construction cost of the dam. In unusual circumstances, the instrumentation cost can be as high as 2 to 3 percent of the total construction cost.

b. *Existing Dams*.—The fact that some existing dams contain only minimal or no instrumentation is not, by itself, considered to be a sufficiently adequate reason for installation of instruments at each and every dam. It is felt that more substantive reasons must be present (i.e., severe downstream hazard, visually noted

---

[1]Numbers in brackets refer to entries in the Bibliography.

2

Table 1-1. – Summary of significant failures of concrete dams.

Name of Dam	Type	Location	Approximate Height, feet (m)	Year of Failure	Approximate Age at Failure, years	Probable Principal Cause
Alla Sella Zerbino	Gravity	Italy	39 (11.9)	1935	12	Foundation seepage, sliding, and overturning
Austin	Gravity	Pennsylvania (U.S.)	50 (15.2)	1911	2	Foundation sliding, concrete cracking
Bouzey	Masonry/ Gravity	France	72 (21.9)	1895	14	Uplift and internal hydrostatic pressures
Dujeprostroj	Gravity	Russia	131 (39.9)	1941	*	Blown up by Soviet troops
Eder	Gravity	Germany	157 (47.8)	1943	29	Bombed during war
Eigiau	Gravity	Wales	35 (10.7)	1925	17	Seepage under dam
Gleno	Multiple-arch and Gravity	Italy	143 (43.6)	1923	0.5	Poor design, poor workmanship
Khadakwasla	Masonry/ Gravity	India	131 (39.9)	1961	82	Uplift pressures, internal cracking
Malpasset	Arch	France	200 (61.0)	1959	5	Weak abutment rock, high water pressures in abutment
Mohne	Gravity	Germany	132 (40.2)	1943	32	Bombed during war
Puentes	Masonry/ Gravity	Spain	164 (50.0)	1802	11	Seepage under dam
St. Francis	Arched/ Gravity	California (U.S.)	205 (62.5)	1928	2	Poor foundation, internal cracking
Tigra	Masonry/ Gravity	India	86 (26.2)	1917	0.25	Overtopping and sliding
Vaiont	Arch	Italy	869 (264.9)	1963	3	Overtopping caused by rockslide in reservoir
Vega de Tera	Buttress, Masonry, and Slab	Spain	112 (34.1)	1959	2	Leakage in construction joints, foundation sliding

*Unknown

problems, etc.) for installing and monitoring instrumentation. This philosophy is believed to have resulted in greater cost effectiveness in the overall Bureau mission. In fact, certain types of instrumentation that must be installed by drilling holes may present a risk to the safety of a facility. The Bureau's goal is to use instrumentation as a major tool in an ongoing commitment to dam safety. Installation of instrumentation at an existing concrete dam is generally a more difficult and costly task than installation of similar instrumentation in an existing embankment dam. Therefore, new instrumentation installation at an existing concrete dam is usually limited to surface devices.

1.5 Minimum Desirable Instrumentation.—In 1979, a team of Bureau personnel was formed to study whether there should be a minimum instrumentation requirement for Bureau dams and, if determined to be so, what should constitute this minimum requirement. At that time, the problem of determining minimum instrumentation for the safety of dams was divided into two categories, existing and new dams, because each category had benefits and problems not common to the other. Some of the important differences noted were:

- New dams would probably have additional monitoring requirements to demonstrate safety during construction, first filling, and early age.

- Additional options for installing instruments were available during construction that could not be considered for existing dams.
- Older existing dams may have already settled or undergone nonrecoverable deformations that were not measurable.
- Many older existing dams were built using design, construction, and performance standards significantly different from existing dams built under more recent state-of-the-art standards.

It was determined that the factors of greatest importance to the safety of various concrete dam types were as follows:

a. *Concrete Arch Dams.*—Because of the monolithic behavior of arch dams, displacement is the most meaningful parameter that can be readily monitored. Although displacements occur in all directions, the most significant displacements are the ones that take place in the horizontal plane. All concrete arch dams should have provisions for measuring these displacements, including relative movements between points within the dam and movement of the dam relative to a remote fixed point. In addition, the quantity and location of all seepage flows, including formed and foundation drains, should be monitored whenever possible. Other parameters that should be considered for monitoring include foundation movement and relative movement at any major joint in the dam or in its foundation that are significant. Water quality should be determined for seepage, reservoir, and tailwaters early in the life of a dam to provide a basis of comparison with later samples.

b. *Concrete Gravity Dams.*—If a gravity or buttress dam maintains its structural integrity and is stable against sliding, no safety hazard is likely to occur. Because contraction joints between blocks are much weaker than the mass concrete, any indication of loss of structural integrity in the dam or foundation will manifest itself at the joints. Instrumentation should be installed in all gravity or buttress dams to monitor relative movement between blocks exhibiting indications of previous movement or where movement might be reasonably anticipated. These measurements of relative movement between blocks should be tied to other measurements that allow displacements of the dam relative to a remotely located fixed point to be determined. Other significant parameters that should be monitored in the majority of gravity or buttress dams are uplift pressures and seepage flows, including foundation drains. Results from these measurements may indicate open joints in the foundation or a crack in the concrete. If there are indications of movement along any crack in the dam or foundation, additional monitoring devices should be installed to measure sliding or rotation along the plane. Water quality should be determined for seepage, reservoir, and tailwaters.

c. *Spillway and Outlet Works.*—Most spillway and/or outlet works structures experience some movement. Although movement, in itself, may not be a positive indication of structural distress, differential movement between structural elements or within different portions of the same structure usually is a definite indication. Means for determining differential horizontal and vertical movements should therefore be installed at vital locations on all existing structures. When measurable seepage is flowing from spillway and outlet drains or adjacent parts of a dam, quantities of flow and water quality data should be obtained.

As a result of the team's efforts and subsequent policy reviews, it was determined that new dams should be designed not only to satisfy a minimum requirement but also to contain modern, high-quality, state-of-the-art instruments to be located at optimum positions, and to the degree required to determine long-time dam safety over a period consistent with engineering design and economic constraints. Existing dams should be examined and retrofitted with instrumentation as dictated by their relative need on a site-specific basis.

1.6 Safety Evaluation of Existing Dams.—In 1978, the Bureau initiated the SEED (Safety Evaluation of Existing Dams) Program along with a comprehensive training program for examining and evaluating the safety of existing dams. A SEED Manual [2] was prepared to provide engineering and technical personnel at all levels of Government and private engineering organizations with sound, comprehensive guidelines and procedures for the examination and evaluation of dams.

A formal inspection team consists of experienced civil and mechanical engineers and geologists, normally led by engineers assigned to the Division of Dam Safety. Other team members may include regional personnel, individual consultants, and other Division of Design staff. Project personnel, representatives from other Federal and State agencies, and water-user group representatives may join the team for the onsite examination.

The team conducts a comprehensive review of all data pertinent to the safety of the dam, makes an onsite examination, analyzes all data and findings, updates a data book, and prepares a written Examination Report to the Chief of the Inspections Branch, Division of Dam Safety. Findings, conclusions, and recommendations relative to the safety of the dam are presented.

B. Hydrostatic Pressure Measuring Devices

1.7 Purpose.—The hydrostatic pressure differential between the reservoir level and the downstream pool or tailwater results in an obvious potential for seepage through, around, and under a dam. Such seepage occurs at every dam through joints or cracks in the dam and through joints, cracks, or bedding planes in the dam foundation and abutment rock.

It is important to measure these water pressures at various points in the dam foundation and abutments because such measurements can be critical to aid in the detection of possible piping or other seepage-induced instability situations, such as the presence of excess hydrostatic uplift pressures on the base of the dam. These measurements also indicate the gradient (upstream to downstream) and provide the means to determine effectiveness of grout curtains and foundation drainage facilities.

1.8 Types.—Many styles and types of hydrostatic pressure measuring devices have been available for use over the years. Those types that will be discussed in detail later in this manual are those which are in use at Bureau facilities. Basically, the general types in use are peizometers, operating either as a closed or open system, and closed system Bourdon-type pressure monitoring systems.

Closed system piezometers in Bureau concrete dams consist of vibrating-wire units or Carlson-type devices, while open system devices used are commonly called observation wells. A variation of the closed system unit is the well or pipe system, which is capped so that a Bourdon-type gauge may be utilized for directly reading water pressure. Some similar systems use pressure transducers rather than Bourdon gauges to measure the pressure.

Other types of piezometers are available for use in concrete dam installations; however, they are not presently being used. These other types include hydrostatic pressure indicators, hydraulic twin-tube piezometers, pneumatic piezometers, porous-tube peizometers, and slotted-pipe peizometers, which are regularly used in Bureau embankment dams and are thoroughly discussed in the Embankment Dam Instrumentation Manual [3]. Advantages and limitations of various piezometer types are shown in table 1-2.

C. Pressure (Stress) Measuring Devices

1.9 Purpose.—In some instances, it is desirable to know the relative stress between a dam and its abutments or foundations, or between components of a dam. Design stresses may or may not actually occur in a completed dam; therefore, certain instruments are used to determine the magnitude of the actual stresses at selected locations.

1.10 Types.—Many types of pressure or stress measuring devices exist; however, the Bureau has used only four types: Gloetzl cell, Carlson load cell, vibrating-wire gauges, and flat jacks. The Gloetzl cell operates hydraulically to balance (null) a given pressure, while the Carlson load cell utilizes changing electrical resistance due to wire length changes caused by applied pressure. The vibrating-wire gauge, a variation of the Carlson cell, measures the change in vibration frequency caused by strain in a vibrating wire. The flat jacks use a Bourdon-tube gauge to measure pressures.

D. Seepage Measurement Devices

1.11 Purpose.—Seepage through or around a concrete dam is an extremely valuable indicator of the condition and continuing level of performance of the dam. The quantity of seepage entering a seepage collection system is usually directly related to the level of the water in the reservoir. Any sudden change in the quantity of seepage collected without a corresponding obvious cause, such as an appropriate change in the reservoir level or heavy rainfall, could signal a seepage problem.

Similarly, should the seepage water become cloudy, discolored, contain increasing quantities of sediment, or change radically in chemical content, a serious seepage problem could likely be indicated. Wet spots or seepage appearing at new or unplanned locations at the abutments or downstream from a dam could also indicate a seepage problem.

1.12 Types.—Commonly used seepage monitoring devices include quantitative devices that include weirs, flowmeters, Parshall flumes, and calibrated catch containers. Geophysical methods used for qualitative seepage

5

Table 1-2. – Advantages and limitations of various piezometer types.

Piezometer Type	Advantages	Limitations
Standpipe piezometer or well point, observation well	Simple, reliable, and long experience record. No elaborate terminal point needed.	Occasionally slow response. Pipe and tubing must be raised nearly vertical. Freezing problems. Subject to damage by construction equipment. Can mean costly drilling and related problems.
Closed hydraulic piezometer	Long experience record, rapid response, and less prone to damage by construction equipment. Can be adapted for continuous recording.	Location of terminal well. Freezing and corrosion problems. Periodic de-airing required. Maintenance problems. Up to 30 percent long-term loss. Complex flushing procedures.
Pneumatic piezometer	Level of terminal independent of tip level, rapid response, and no freezing problems. Can be adapted for continuous recording.	Must prevent humid air from entering tubing. Shorter experience record than hydraulic piezometer. Skilled operator needed.
Vibrating-wire piezometer	Simple to read and maintain, level of terminal independent of tip level, rapid response, and high sensitivity. Suitable for automatic readout or data logger. Frequency signal permits data transmission over long distances or to existing structure. No freezing problems. Can be used to read negative pore pressures.	Sensitive to temperature. Slight risk of zero drift. Sensitive to barometric changes. Longevity unknown.
Resistance strain gauge piezometer	Level of terminal independent of tip level, rapid response, and high sensitivity. Suitable for automatic readout. No freezing problems. Can be used to read negative pore pressures.	Sensitive to temperature. Risk of zero drift, sensitive to moisture, cable length, change in connections.

analysis include thermotic surveys and self-potential measurements. To date, the use of geophysical methods by the Bureau has involved only embankment dams.

Weirs are one of the oldest, simplest, and most reliable devices that can be used to measure quantity flow of water. The critical parts of weirs can be easily inspected, and any improper operation can be easily detected and quickly corrected. Types of weirs normally used are the 90° V-notch, rectangular, or trapezoidal (Cipolletti). The quantity discharge rates are determined by measuring the vertical distance from the crest of the overflow portion of the weir to the water surface in the pool upstream from the crest. The discharge may then be computed by an appropriate formula or by reference to tables prepared for that purpose.

A Parshall flume is a specially shaped open-channel flow section. The discharge may be computed or determined by reference to table and charts prepared using throat width of flume, upstream head, and downstream head as variables.

Calibrated containers may be used to measure low flow quantities from a pipe outlet. The time required to fill a container of known volume is measured and then the flow is computed, usually in gallons per minute (cubic meters per second).

Flowmeters and pressure transducer devices are also sometimes used to determine quantity of flow in a pipe or open channel.

Thermotic survey techniques may, in special instances, aid in identifying zones of high permeability and ground-water flow concentrations within fractured rock and alluvial deposits. Although these techniques do not replace the need for borings or the need to install conventional instrumentation, they may be valuable in directing the location of more quantitative investigation methods such as drill holes and pumping tests.

Self-potential or streaming potential surveys may also be useful in the detection of discrete seepage paths that tend to render conventional piezometer data inadequate.

E. Internal Movement Measuring Devices

1.13 Purpose.—Measured internal movements of dams may be of importance in any plane or orientation, but generally consist of essentially horizontal and vertical measurements. It is usually desired to obtain measurements of relative movements between portions of dams and between dams and their abutments and/or foundations. Therefore, relative movements are usually observed along joints or cracks, between sections of appurtenant structures such as penstocks, and internal deformations of structures during cyclic loading conditions of temperature and water.

1.14 Types.—Many devices are available for measuring internal movements. Those devices currently in use or being considered for use by the Bureau include calibrated tapes, single-point and multipoint borehole extensometers, joint meters, plumblines, dial gauge devices, Whittemore gauges, resistance gauges, tilt meters, and inclinometer/deflectometers. Strain meters and "no-stress" strain devices may also be used for measuring internal movements. The devices may be installed within the interior of a dam, in the foundation or abutments, on the dam surface, in inspection galleries, or on penstocks or powerplants.

F. Surface Movement Measuring Devices

1.15 Purpose.—External vertical and horizontal movements are measured on the surface of concrete dams to determine total movements with respect to a fixed datum located off the dam. Reference points may be monuments or designated points on the crest, on the upstream and downstream faces, at toe of dam, or on appurtenant structures. Both lateral or translational and rotational movements of the dam are of interest.

1.16 Types.—Surface movements are usually observed using conventional level and position surveys. The position surveys may be conducted using triangulation, trilateration, or collimation techniques. Individual measurement devices include levels, theodolites, calibrated survey tapes, EDM (electronic distance measuring) devices, and associated rods, targets, etc.

G. Vibration Measuring Devices

1.17 Purpose.—Major vibrations at a damsite could result in concrete cracking and/or liquefaction and potential stability problems of appurtenant embankments. These vibrations could be caused by naturally occurring earthquakes or by construction (rehabilitation) related vibrations from blasting.

Measurement of earthquake motion is necessary to improve the design of dams so that they can better resist earthquake effects and to asssit in damage assessment following significant earthquake occurrences. Measurement of construction-induced vibration is needed to control construction activity in the vicinity of a dam. Because of the inability to predict exactly when or where an earthquake will occur, it is desirable to instrument most structures in areas of high earthquake incidence and other structures where appurtenant embankments or foundation materials may potentially liquify during seismic action. The U.S. Geological Survey currently installs and reads all permanently installed seismic devices for the Bureau.

1.18 Types.—Vibration or seismic instrumentation is used to record responses of the structure, foundation, and abutments to seismic events. The general term "seismograph" refers to all types of seismic instruments that automatically "write" a permanent, continuous record of earth motion. The basic components of a seismograph include a frame anchored to the ground or dam, one or more transducers, timing devices, and a recorder. As the frame moves with the dam, the transducers respond according to principles of dynamic equilibrium. Signals of horizontal motion in two planes and vertical motion may be sensed either electrically, optically, or mechanically; and the motion sensed may be proportional to acceleration, velocity, or ground displacement.

Various commercially available instruments include the strong motion accelerograph, peak recording accelerograph, and others.

7

H. Monitoring Schedules

1.19 General.—Instruments are installed at a dam to assess the performance and safety of the structure during construction and over the long term. Obviously, periodic readings of the instrumentation are required to accomplish that goal. An analysis of these readings can indicate unusual behavior of a structure.

During filling of the reservoir, uplift pressure pipes should indicate if a pattern of seepage flow exists through the abutments and foundation. Observations made immediately following a severe drawdown of the reservoir will provide information on pore water pressures retained within and under the structure. Changes in pore water pressures and internal or external movement may reflect stress conditions within the dam that influence the stability of the structure. Because behavior conditions originate during construction and vary continuously throughout operation, instrumentation data may be utilized by construction and operating personnel as well as by design engineers.

The processing of voluminous raw data is efficiently handled using current computer technology. The interpretation of the data requires careful examination of measurements as well as other influencing effects such as reservoir operation, air temperature, precipitation data, seasonal shutdown during construction, and periodic instrument operation evaluations.

Guidelines for the frequency of readings on typical instrumentation systems are shown in table 1-3.

1.20 Variations in Schedule.—Each dam should have a prescribed monitoring schedule for each phase in the life of the dam such as construction, construction shutdown, first reservoir filling, and long-term operation including drawdown periods. In addition, variations in the schedule may be required for special conditions that may arise during the life of a dam such as noticeable or unusual changes in readings, seismic activity, and rehabilitation construction work. Most variations in schedule will result in a temporarily increased frequency of readings.

1.21 Emergency Procedures.—In situations where the safety of a dam is compromised and a failure or impending failure condition is present, it is extremely important that dam personnel immediately follow exact prescribed procedures. Each dam has such a set of prescribed procedures listed in the SOP (Standing Operating Procedures) for the dam. These procedures cover such items as earthquakes, observed cracking, new springs or seeps, changes in quantity and coloration of seepage water, landslides, and abnormal instrument readings. Proper compliance with emergency procedures requires that the instrumentation readers be diligent, alert, and very conscientious in reporting any unusual situation observed.

I. Maintenance And Performance

1.22 General.—All instrumentation such as piezometers, stress and strain meters, and inclinometers; and accessory equipment such as cables, tips, valves, and gauges should be inspected and/or calibrated over their expected operating range as soon as possible after delivery from the suppliers so that defects can be corrected before installation. Removable items of equipment may be recalibrated at regular time intervals.

To properly monitor the performance of a dam, it is necessary to collect instrumentation data over extended periods. Therefore, it is important that the monitoring equipment be as simple, rugged, and durable as possible and be maintained in satisfactory operating condition. The necessity of maintaining proper operational characteristics creates many problems. Even a simple surface leveling point may be subject to damage by frost action, traffic and maintenance operations on the crest, or vandalism. Observation wells and most piezometers can be damaged by frost action, caving, corrosion of materials used for casing, loss of measuring equipment in hole, and by vandals dropping rocks into the holes. Unless special precautions are taken, the average life of installations of these types may be significantly reduced.

To minimize damage, the tops of measuring points and wells should be as inconspicuous and close to the surrounding surface as possible. Locations of installations should not be immediately adjacent to roads, trails, or water channels; and noncorrosive materials should be used wherever possible.

The problems of providing for satisfactory continuing performance are significantly greater for any apparatus that is permanently embedded in the dam, dam foundation, or abutments. Some instruments may be installed with the knowledge that they will provide information until their useful life is over or until readings have provided enough information to satisfy design or performance criteria.

8

Table 1-3—Suggested minimum frequency of readings[1].

Type of Instrument	During Construction		During Initial Filling	Periodic Report of Operations		
	Construction	Shutdown		First Year	2 to 3 Years	Regular
Vibrating-wire piezometers	Weekly[2]	Monthly	Weekly	Biweekly	Monthly	Monthly
Hydrostatic uplift pressure pipes	Weekly	Monthly	Weekly	Weekly	Biweekly	Monthly
Porous-tube piezometers	Monthly	Monthly	Weekly	Weekly	Monthly	Monthly
Slotted-pipe piezometers	Weekly	Monthly	Weekly	Weekly	Biweekly	Monthly
Observation wells	Weekly	Monthly	Weekly	Weekly	Biweekly	Monthly
Water levels	Weekly	Monthly	Weekly	Weekly	Monthly	Monthly
Seepage measurement (weirs. flumes. etc.)	—	—	Weekly	Weekly	Biweekly	Monthly
Visual seepage monitoring	Weekly	Weekly	Weekly	Weekly	Biweekly	Monthly
Resistance thermometers	Twice weekly	Monthly	Weekly	Weekly	Monthly	Monthly
Thermocouples	Daily	Monthly	Weekly	Weekly	Monthly	Monthly
Carlson strain meters	Weekly	Weekly	Weekly	Biweekly	Monthly	Monthly[3]
Joint meters	Weekly	Weekly	Weekly	Biweekly	Monthly	Monthly[3]
Stress meters	Weekly	Monthly	Weekly	Biweekly	Monthly	Monthly[3]
Reinforcement meters	Weekly	Monthly	Monthly	Monthly	Monthly	Monthly[3]
Penstock meters	Weekly	Monthly	Monthly	Monthly	Monthly	Monthly[3]
Deflectometers	Weekly	Monthly	Weekly	Weekly	Monthly	Monthly
Vibrating-wire strain gauge	Weekly	Monthly	Monthly	Monthly	Monthly	Monthly
Vibrating-wire total pressure cell	Weekly	Monthly	Monthly	Monthly	Monthly	Monthly
Load cell	Weekly	Monthly	Weekly	Biweekly	Monthly	Monthly
Pore pressure meters	Weekly	Weekly	Weekly	Biweekly	Monthly	Monthly
No-stress strain meters	Weekly	Weekly	Weekly	Biweekly	Monthly	Monthly
Foundation deformation meters	Weekly	Weekly	Weekly	Biweekly	Monthly	Monthly
Flatjacks	Daily	Weekly	Weekly	Biweekly	Monthly	Monthly
Tape gauges (tunnel)	Weekly	Weekly	Weekly	Biweekly	Monthly	Monthly
Whittemore gauges	—	—	Biweekly	Biweekly	Monthly	Monthly
Avongard crack monitor	Weekly	Monthly	Weekly	Weekly	Monthly	Monthly
Wire gauges	—	—	Monthly	Monthly	Monthly	Quarterly
Abutment deformation gauges	Weekly	Monthly	Weekly	Weekly	Monthly	Monthly
Ames dial meters	Weekly	Monthly	Weekly	Weekly	Monthly	Monthly
Differential buttress gauges	Weekly	Monthly	Weekly	Weekly	Monthly	Monthly
Plumblines	Daily	Weekly	Daily	Weekly	Biweekly	Monthly
Inclinometer	Weekly	Weekly	Weekly	Weekly	Biweekly	Monthly
Collimation	Every other day for a month	Monthly	Weekly	Biweekly	Monthly	Monthly
Embankment settlement points	—	—	Monthly	Bimonthly or six times per year	Quarterly	Two times per year
Level points	Monthly	Quarterly	Monthly	Monthly	Six times per year	Six times per year
Multipoint extensometers	Weekly	Monthly	Weekly	Monthly	Monthly	Quarterly
Triangulation	—	—	Monthly	Monthly	Quarterly	Two times per year
Trilateration (EDM)	—	—	Biweekly	Monthly	Quarterly	Quarterly
Reservoir slide monitoring systems	—	—	Monthly	Monthly	Monthly	Quarterly
Powerplant movement	—	—	Weekly	Monthly	Monthly	Monthly
Rock movement	Weekly	Monthly	Weekly	Monthly	Monthly	Monthly

[1] These are suggested minimums; however. anomalies or unusual occurrences such as earthquakes or flood will require additional readings.
[2] Daily during curtain grouting.
[3] May be discontinued after 3 years unless anomalies are noted.

If certain types of piezometer tubing are used, there are certain microbes that can live and proliferate within the tubes unless the water in the system is treated with a biological inhibitor. Some antifreeze solutions previously placed in systems develop a floc that results in plugging of the tubes. Also, in certain environments, materials in some gauges may corrode and render them useless.

As engineers have become aware of the problems associated with the usage of various materials and instrumentation devices along with the knowledge of the environment in which they are used, modifications have been made in the design, installation procedures, and operating instructions to ameliorate these difficulties.

1.23 Performance Checks.—Many devices such as gauges and various reading devices are removable and may be calibrated on a regular basis. However, most of the instrumentation at a dam is fixed in place and not repairable when damage or malfunctioning is discovered. Fixed devices can generally only be replaced from the surface by devices installed in drilled holes and are, therefore, usually not replaceable. Other devices, such as surface monuments, are replaceable to some extent.

1.24 Calibration of Devices.—On request from a regional or project office or on the assessment of the Structural Behavior Branch Reviewer, the Structural Behavior Branch will assist in the replacement or recalibration of defective or suspected faulty equipment. Pressure gauges for uplift pipes are routinely calibrated prior to field use. Inclinometers or deflectometer systems (probes and readout devices) are checked in a hole of known inclination before field use and periodically rechecked. As part of the purchasing process for new instruments (gauges, readout units, and piezometers), a manufacturer's certificate and calibration data are required. This data ensures that gauge factors and other important details of the instrument are known before preinstallation testing in the field begins.

All piezometers should be tested for proper response to a known pressure before installation. These tests should include examination for air leaks on pneumatic piezometers, checking of the gauge factor supplied by the manufacturer of a vibrating-wire instrument, verification that factory saturation of ceramic stones for proper air removal was done, and checking for leaks in total pressure cell plates. These tests are usually performed at the jobsite but in some cases have been performed at the Bureau's E&R (Engineering and Research) Center.

J. Data Acquisition, Processing, and Review Procedure

1.25 General.—Acquisition and processing of a vast amount of data are necessary to maintain effective monitoring of Bureau concrete dams. Currently, 50 dams are being monitored to varying degrees resulting in more than 1 million data items annually handled and processed by the Concrete Dam Instrumentation Section of the Structural Behavior Branch. The specific goals of the collection, processing, and review procedures are to provide accurate and timely evaluation of data relating to the safety of the facility. The continuing process whereby instrumentation data are acquired, compiled, and appropriate action taken, based on the data, may be grouped into five phases: (1) data acquisition, (2) transmittal, (3) processing, (4) review, and (5) action. These phases are illustrated on figure 1-1.

    a. Data Acquisition.—Instrumentation data are normally obtained from instrument readings and observations by Bureau project personnel. In some instances, the water user or contractor personnel obtain the data. The data are acquired in accordance with a specific monitoring schedule that has been established for each instrument type at each dam. An example of a typical Concrete Dam Instrumentation Reading Schedule form is shown on figure 1-2.

In most cases, instrumentation data are recorded manually on a field data form for transmittal to the Structural Behavior Branch. Several examples of field data forms are presented later in this manual. A relatively recent development in field data acquisition involves the use of portable teletranslational devices, as illustrated on figure 1-3. Currently, data acquisition personnel are evaluating this equipment at several dams. Using this type of device, the instrumentation reader enters the field data directly into the device's memory. A previous reading for each instrument may be stored in the device and displayed upon request. Comparison of a previous reading with the present reading can be very helpful in eliminating gross reading errors and data entry errors. Upon completion of the instrument readings, the data in the device are transmitted to a computer terminal for forwarding to the Structural Behavior Branch in Denver. A telephone modem may be used when a terminal is not available. When data are transmitted by computer, the transmitting office may, upon request, receive

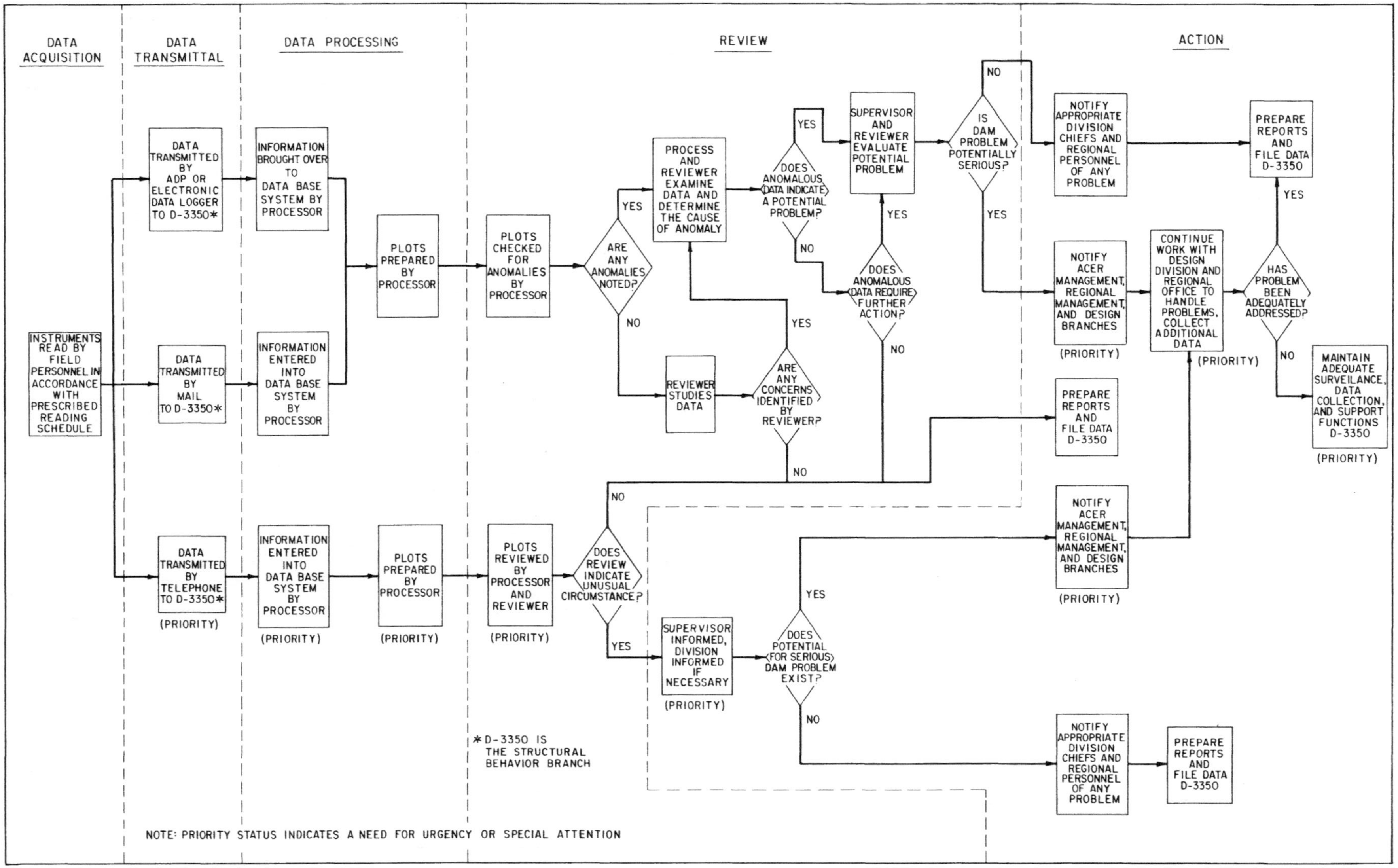

Figure 1-1.—Data acquisition and processing phases. 40-D-6514C.

NOTE: PRIORITY STATUS INDICATES A NEED FOR URGENCY OR SPECIAL ATTENTION

*D-3350 IS
THE STRUCTURAL
BEHAVIOR BRANCH

11

CONCRETE DAM INSTRUMENTATION
READING SCHEDULE

Dam_____Date_____
Project_____Prepared by_____

	INSTALLATION	SCHEDULE	REFERENCE
Horizontal Deflections	Collimation		
	Plumblines		
	Other (EDM, etc.)		
	Uplift Pressures		
	Drain flows/Seepage		
	Embedded 'Carlson' Type Instruments		
	Other Instruments (Inclinometers, MPBX, SPBX, Strain Gages)		

Remarks:

Figure 1-2.—Typical form for an instrumentation reading schedule.

12

Figure 1-3.—Portable teletranslational device. From Telxon Portable Tele-Transaction Computer Operator's Guide. P801–D–81188.

a listing of the data sent and a comparison to the last set of data. Any required mathematical computations are performed by the computer program. Computer entry should result in more rapid, timely, and accurate data acquisition and transmittal.

b. ***Data Transmittal.***—All instrumentation data are transmitted to the Structural Behavior Branch (code D-3350) at the E&R Center in Denver, Colorado. Data may be sent by mail or transmitted directly by the CCS (Central Computer System) using remote terminals and telephone links. Specially requested data for dams may be transmitted verbally by telephone, when computer communications is not available, on a frequent schedule (even daily) to the Structural Behavior Branch. Also, a few dams are scheduled to have their data transmitted automatically via GOES (Geostationary Operational Environmental Satellite) directly to the E&R Center.

c. ***Data Processing.***—Regardless of the mode of data transmittal, all data are entered into the Structural Behavior Branch Hewlett-Packard computer system for processing. Data received by the CYBER system are processed within 24 hours from time entered. Data received by mail are also usually processed within 24 hours. Data received by telephone are processed and reviewed within a few hours of receipt.

Utilizing the computer system, data are always compared with previous data using computer routines to prepare graphic presentations of the data, such as current reservoir level. Other graphs may plot variables on each axis; e.g., seepage versus reservoir levels to correlate cause and effect. Examples of these plots are included in succeeding chapters. The generated plots are delivered to a designated reviewer in the Concrete Dams Instrumentation Section who is experienced and knowledgeable in the field of instrumentation and concrete dam behavior, see figure 1-1. All data received by mail are logged into the system by one person, thereby automatically updating a status sheet maintained for all dams. This status sheet is reviewed continually for late or missing data.

d. ***Data Review.***—Initially, the technician who enters the data reviews the input for obvious anomalies and ensures, as is possible, that the information is correct. Then, the designated reviewer for each dam checks each plot for any anomalies or unusual trends in the recently acquired data. Such anomalies may include unusual readings or may simply represent a change in data trends which, at least initially, may not conform with previous trends. However, the majority of anomalies are the result of human error in the acquiring, transmitting, or entering of the data into the computer. Potential problems that have been detected by the instrumentation system are discovered during this phase of the process.

The computer plots provide a comparison of the new data with all previous data from the same instruments. The obvious data errors are usually quickly rectified by contacting the person that made the reading and requesting a supplementary reading. If the data are correct and a potentially serious problem is detected, the reviewer's supervisor is consulted and a decision made on the proper course of action.

One copy of the data plots is prepared by the computer system. After review, the plots are placed in a book maintained in a current status for reference use by various personnel at the Bureau's E&R Center in Denver.

Upon request, a copy of the data plots will be provided to interested agencies, such as the field office responsible for acquisition of the data. It is felt that full and complete communication of the processed data to regional and field offices is critical to the team effort required by the Bureau's instrumentation program.

e. Possible Actions.—If no unusual circumstances or potential problems are detected in the data itself or in the potential consequences of the data, the data is filed for future reference. These data are used for preparation of Structural Behavior Reports on each dam. These formal reports are updated every other year.

Should the instrumentation data indicate a serious problem at a dam, the regional and ACER (Assistant Commissioner, Engineering and Research) management personnel are notified of the situation and appropriate action taken. Subsequent meetings at the management level are conducted to determine the proper course of action, which might include such drastic measures as emergency lowering of the reservoir. Regardless of the management decisions made, it is normal to continue monitoring at an increased frequency until such time as the potential problems are considered to have abated or are no longer significant.

All five phases of the process of data management, review, etc., are obviously of great importance in dam safety. However, it must be stressed that the latter four phases are not of any significance unless the initial data acquisition phase is conducted in an accurate and timely manner by personnel who possess a sincere interest in the future safety of the facilities. The personnel collecting the data have the key role in the process and should be appropriately trained because they are in the ideal position to raise immediate concerns if significant anomalies occur.

1.26 Visual Inspections.—The visual inspection of the condition of a dam and the surrounding area by personnel familiar with that specific dam may be the most important factor in the continuing safety of the dam. The previously discussed SEED Program provides for regular formal inspections including visual inspections. The required visual inspections are detailed in appendix B of reference [2]. An extremely valuable supplement to the SEED Program is the performance of visual inspections by the instrument readers at the regular instrument reading intervals.

None of the instrumentation described in this manual will detect an impending failure if the instrumentation is not appropriately located or is, for some reason, not functioning properly. Visual inspection then becomes a most vital factor. Frequently, a visual inspection appropriately reported on the field data forms may serve to explain the reasons for anomalies in the instrument readings. A properly conducted visual inspection necessarily includes walking along the crest, both abutments, and for some distance both upstream and downstream of the dam, if possible. The principal items to watch for include:

- Cracking of the dam in any plane or direction
- Cracking or landslide-type movements of the upstream and downstream valley walls
- Bulging or sloughing of lower portion of any appurtenant embankment slopes, abutments, or valley walls
- Subsidence of any portion of the crest of embankment sections
- Sinkholes in reservoir bottom
- New seeps or springs, or an increase in volume from an existing spring in abutments, downstream valley walls, or floor
- A persistent vortex (whirlpool) in reservoir that is unrelated to any operational outlet works
- Seepage water that is discolored or carrying soil or rock sediment
- Cracking in any of the concrete appurtenant structures
- Offsets in joints in the concrete dam

If any of these items or any other physical occurrence that is not readily explainable is noted, the project office and appropriate personnel at the E&R Center should be contacted as soon as possible. It is also desirable to obtain photographs at regular intervals of any distress noted.

The frequency of visual inspections will vary greatly depending upon the dam characteristics, its age, reservoir level with respect to previous maximum levels, and many other factors. In general, all maintenance and operations personnel at a dam should be constantly alert for signs of distress. The individual at a damsite responsible for instrumentation should perform visual inspections at least at the time of instrument readings. Obviously, many dams are located in areas where access is limited during the winter months and many of the distress features may be obscured by snow. At those locations, formal visual observation should be conducted as soon as possible after snowmelt.

Chapter 2

PRESSURE MEASURING DEVICES

A. Hydrostatic Pressure Measuring Devices (Open System)

2.1 General.—Hydrostatic pressure differentials exist between the reservoir level and the downstream tailwater. These differentials are characterized by seepage through and around a dam. The water pressures are measured at various points along the seepage paths to detect any seepage-induced instability situations such as the presence of excess hydrostatic uplift pressures on the base of the dam. Normally, open- or closed-type system piezometers are used to measure these water pressures. Although most concrete dams are constructed in relatively narrow canyons, some are located in broader valleys that necessitate embankments on either end. The following discussion on hydrostatic pressure measuring devices applies mainly to these embankment portions of dams.

Typical open-system piezometers in use include simple observation wells, porous-tube piezometers, and slotted-pipe piezometers; and closed-system devices include vibrating-wire piezometers and uplift monitoring systems consisting of a well, drain, or an uplift pressure measuring system with an attached Bourdon-type gauge. Closed system devices are discussed in part B of this chapter.

2.2 Observation Wells.—a. *Usage.*—Observation wells are installed in the foundations or abutments of dams, and are used to measure average water level elevations in different zones of materials or over long zones in drill holes. These wells allow water to pass through a slotted or perforated pipe or rise up through the bottom of an open pipe. The water that is being measured may come from different zones and is not isolated by seals in the hole; however, multiple pipes may be installed in a single hole to monitor various zones. Thus, an average piezometric level is measured. The wells may be installed in areas where the piezometric level of a certain zone is needed or may be used to measure the average ground-water elevation in the abutments or downstream foundation. Additional uses of observation wells in conjunction with Bourdon-type pressure gauges are discussed in section 2.6.

b. *Advantages and Limitations.*—Observation wells are relatively easy to install and read, and are economical. A particular disadvantage of these wells is that a perched water table or artesian pressure can occur in a specific stratum that may be interconnected to other strata by the drill hole. The water level observed would then only represent the average piezometric pressure existing throughout the depth of the well. Observation wells are generally installed in each aquifer so that a base reference corresponding to equilibrium conditions can be determined. Table 1-2 in section 1.8 lists other advantages and limitations.

c. *Description of Devices.*—Observation well systems consist of an inlet pipe, standpipe, water-level indicator, protective cover pipe, and backfill materials. A typical installation is shown on figure 2-1.

(1) *Inlet Pipe.*—The inlet or well pipe may consist of virtually any size pipe constructed of any stable material; however, it is usually a 1.5-inch (38-mm) inside diameter, schedule 80, PVC pipe. The pipe is usually provided in either 5- or 10-foot (1.5- or 3.0-m) sections, and is slotted or perforated for water inlet purposes. If slotted, the slots are sawed to a 0.020-inch (0.51-mm) width and spaced at ¼-inch (6-mm) centers in three rows 120° apart around the circumference of the pipe. The pipe sections are provided with internal threads on one end and external threads on the other end. A plug is usually placed in the bottom end of the pipe.

(2) *Standpipe.*—The standpipe extends from the top of the inlet pipe section to the surface at the top of the hole, or to the elevation from which measurements will be made. This pipe is also usually a 1.5-inch i.d., PVC pipe; however, it is not slotted. If it is desired to place more than one observation well per hole, a smaller diameter standpipe and pipe reducer may be used. For this case, the inlet pipes must be isolated from each other by using a bentonite seal between their elevations.

(3) *Backfill Materials.*—Typical backfill materials consist of graded sand or concrete sand placed around the inlet pipe, and sand or bentonite pellets placed around the standpipe.

(4) *Water-Level Indicator.*—The water-level elevation in the standpipe is determined using a water-level indicator (fig. 2-2). This indicator consists of a self-contained reading unit and a probe, and operates

17

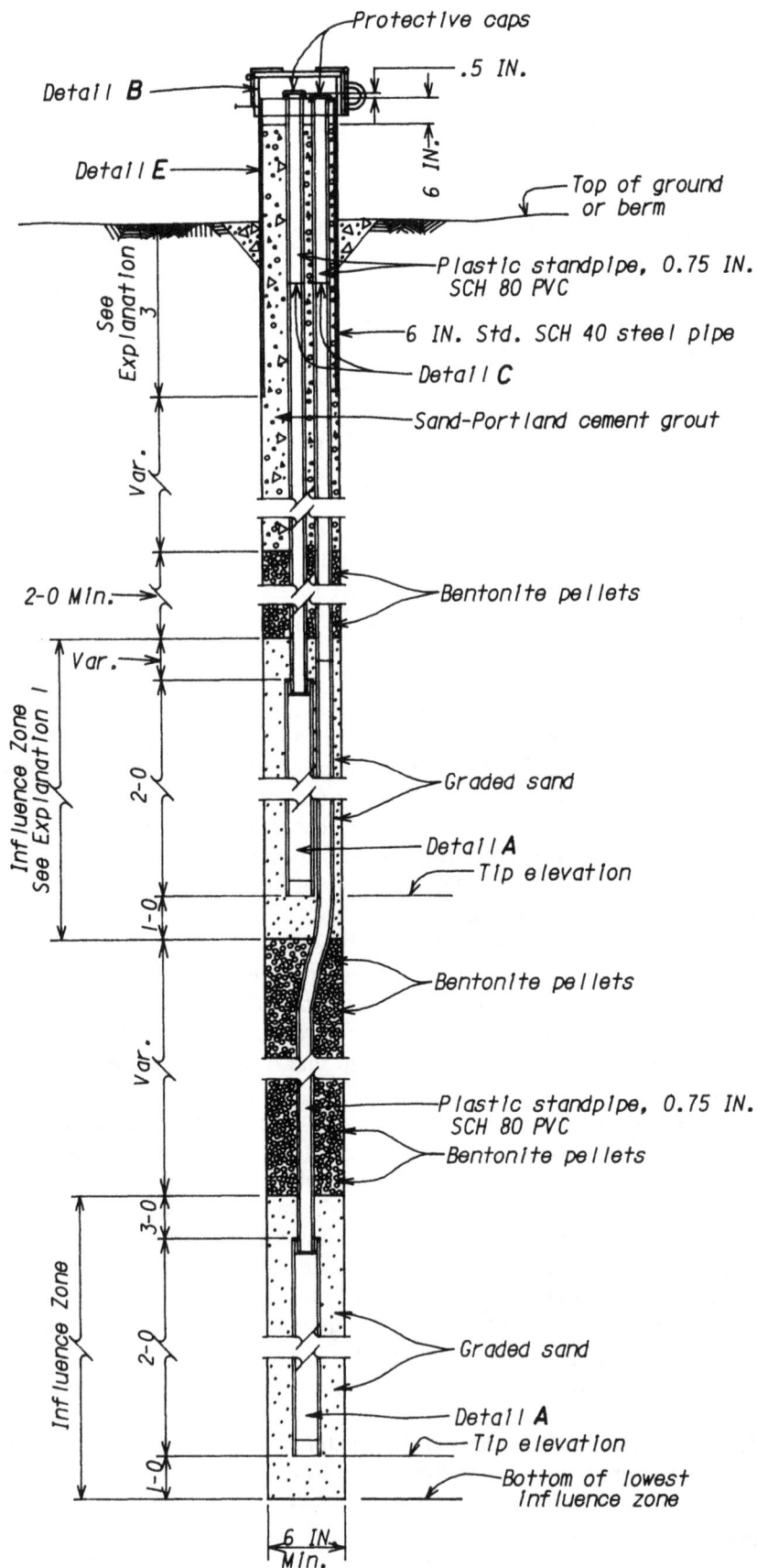

Figure 2-1.—Typical observation well installation. From 596–D–529.

Figure 2-2.—Typical water-level indicator units. Slope Indicator Co. P801-D-81189.

electronically from batteries. When the probe encounters water, an electrical circuit is closed and is indicated by a dial, light, or a sound. At that point, the depth to the water is read from the marked wire connecting the reading unit and probe. Water-level elevations are normally read to the nearest 0.01 foot (0.003 m). The actual water-level elevation is determined by subtracting the depth reading from the previously determined elevation at top of the standpipe.

(5) *Protective Cover Pipe.*—If the standpipe reaches the surface outside of the dam proper, a protective cover pipe is required for protection from vandalism and general security. This cover pipe should have a 4-inch (102-mm) minimum diameter and be at least 6 feet (1.8 m) long. The pipe should also be schedule 40 galvanized iron, and should be fitted with a protective cap and locking device. If the standpipe surfaces within an observation gallery within the dam, only the fitted cap is required.

d. Installation Procedures.—Observation wells may be installed in drill holes in the dam foundation or abutments. The recommended procedure for drill hole installation is similar to that used for observation well installation in embankment dams.

The drill hole for a single observation well should be 4-inch (102-mm) minimum diameter to allow room for sand backfill. The hole may be advanced by either hollow-stem auger or driven casing if in soils, and by either tri-cone drilling or diamond-bit coring in rock strata.

The slotted or perforated inlet pipe with attached standpipe is lowered into the hole and sand backfill placed around it to at least 2 feet (0.6 m) above the top of the inlet section. The balance of the hole may be backfilled using either sand-cement grout or a sand-bentonite mixture. When the hole is backfilled to within about 3 feet (0.9 m) of final surface, the protective cover pipe is placed. This pipe is then surrounded by a compacted clayey soil mixture obtained at the site or by mixing bentonite with soil to produce a low permeability material. The cover pipe should extend 3 feet below ground or to a depth just below the maximum frost line in the area. Standpipes extending into the dam and terminating in an observation gallery or at the surface of a concrete dam have their backfill only to the base of the concrete dam, and are surrounded by concrete above that level.

e. Monitoring Procedures.—Observation wells are monitored using a water-level indicator as follows:
Reading unit's indicator switch is turned on and, if available on instrument, switched to "battery check" position to determine level of charge in batteries. If a satisfactory level of charge is present, proceed:

1. Proper operation of unit may be checked by immersing probe in water and noting if device registers a closed circuit by voltmeter, light, or sound modes.
2. Observation well protective cap is then removed and probe lowered into standpipe. At the level where

19

circuit is closed, note length of probe wire required to reach from water level to top of standpipe to an accuracy of 0.01 foot (0.003 m).

3. Data and reservoir and tailwater levels at time of observation are recorded on a standard data form as shown on figure 2-3. Elevation of top of standpipe has previously been determined.

4. Replace protective cap and lock, if required.

f. Data Acquisition and Processing.—Readings are obtained in feet or meters of depth from top of standpipe. The depth to the water level is subtracted from the elevation at top of standpipe, which yields the water-level elevation.

The data are then transmitted to the Denver Office where the data are processed, reviewed, and action taken as indicated in section 1.25 and on figure 1-1. Example data plots are shown on figure 2-4.

2.3 Slotted-Pipe Piezometers.—Slotted-pipe piezometer systems are nearly identical to the previously described observation well systems (sec. 2.2) in materials, installation, monitoring techniques, and data processing. The only significant difference is that slotted-pipe piezometers are designed and installed to monitor the hydrostatic water pressures in a single stratum or a small zone of a stratum. Thus, the installation always includes bentonite seals above (and sometimes below) the inlet pipe section so that the zone in question is isolated from all other water-bearing zones. A minimum of 5 feet (1.5 m) of bentonite seal is usually specified to ensure zone isolation. A typical data plot illustrating slotted-pipe piezometer data is shown on figure 2-5.

2.4 Porous-Tube Piezometers.—a. *Usage.*—Porous-tube piezometers are used to measure water pressures in a dam foundation or abutment. A drill hole installation procedure is normally used for this type of piezometer. The selection of a porous-tube piezometer over a slotted-pipe piezometer is usually directly related to the grain size of the materials at the measuring piezometer tip (the influence zone). Fine-grained materials around the piezometer tip could plug a slotted-pipe piezometer. Porous-tube piezometers are also used at locations where it is desired to measure the water pressure in a relatively small zone of materials and as check measurement devices for closed-system piezometers installed nearby.

b. Advantages and Limitations.—Advantages of porous-tube piezometers include relatively inexpensive installation (unless drilling costs are high), simple operation, no metallic parts to corrode, and they have a very reliable, long-performance record. When compared to a conventional slotted-pipe observation well, porous-tube piezometers have the following distinct advantages:

- The large intake area of the tube relative to the small diameter of the standpipe tends to minimize the quantity of water required for equalization of pore pressure changes.
- The permeable intake space can be positioned to isolate water pressures occurring in a stratum of limited thickness, even though such pressures are in excess of normal hydrostatic pressure. If excess pressure should raise the water level above the top of the standpipe, additional riser standpipe sections may be added or a hydrostatic Bourdon-type gauge can be installed.
- The installation is fabricated from durable and inert materials that are unaffected by deterioration or corrosion.
- Basic tests can be performed on the piezometer after installation to appraise the sensitivity and to determine the average permeability of the material surrounding the tip.

Limitations on the use of porous-tube piezometers include the fact that the filter media surrounding the porous tip can become plugged due to repeated water inflow and outflow, the standpipe must extend nearly vertically, shallow installations may freeze, and alterations must be made if the piezometric level is above the top of the standpipe. The potential plugging problem occurs in installations with clay- or silt-sized particles at the tip elevation. These particles tend to penetrate into the sand backfill, which reduces the sensitivity of the piezometer.

The sand backfill that completely surrounds the porous tube should meet standard filter requirements as closely as possible without including any silt-sized particles. The porosity of the porous tube may be selected from coarse, medium, or fine grades to inhibit the movement of particles from the surrounding soil through the tube.

Experience has indicated that the performance of a porous tube is not entirely satisfactory when used in materials containing an appreciable air content. The presence of air in the soil or rock void spaces will decrease the average soil permeability from that which is present during a completely saturated condition.

20

OBSERVATION WELL READINGS

MODIFIED_____FOR

Dam_____Date of Observations_____

Project_____Observer_____Sheet_____of_____

Ref. Dwg._____ Res. Water elev. (1)_____Tallwater elev. (1)_____

WELL NO.	LOCATION (2)		ORIGINAL ELEV. (3)	ELEV. - TOP RISER TUBE (4)		SETTLE-MENT-TOP RISER TUBE *	DISTANCE-TOP RISER TUBE TO WATER SURFACE	ELEV. WATER IN WELL
	STATION	OFFSET		ORIGINAL	PRESENT			

(1) Record if approplate. (2) Record distance U/S or D/S from
₵ Crest of dam, or location by coordinates. (3) Taken as
bottom of legth of slotted pipe. (3) Record all elevations
and distances at 0.0Ifoot. *Use minus (-) to indicate heave.

Figure 2-3.—Field data form for observation wells.

21

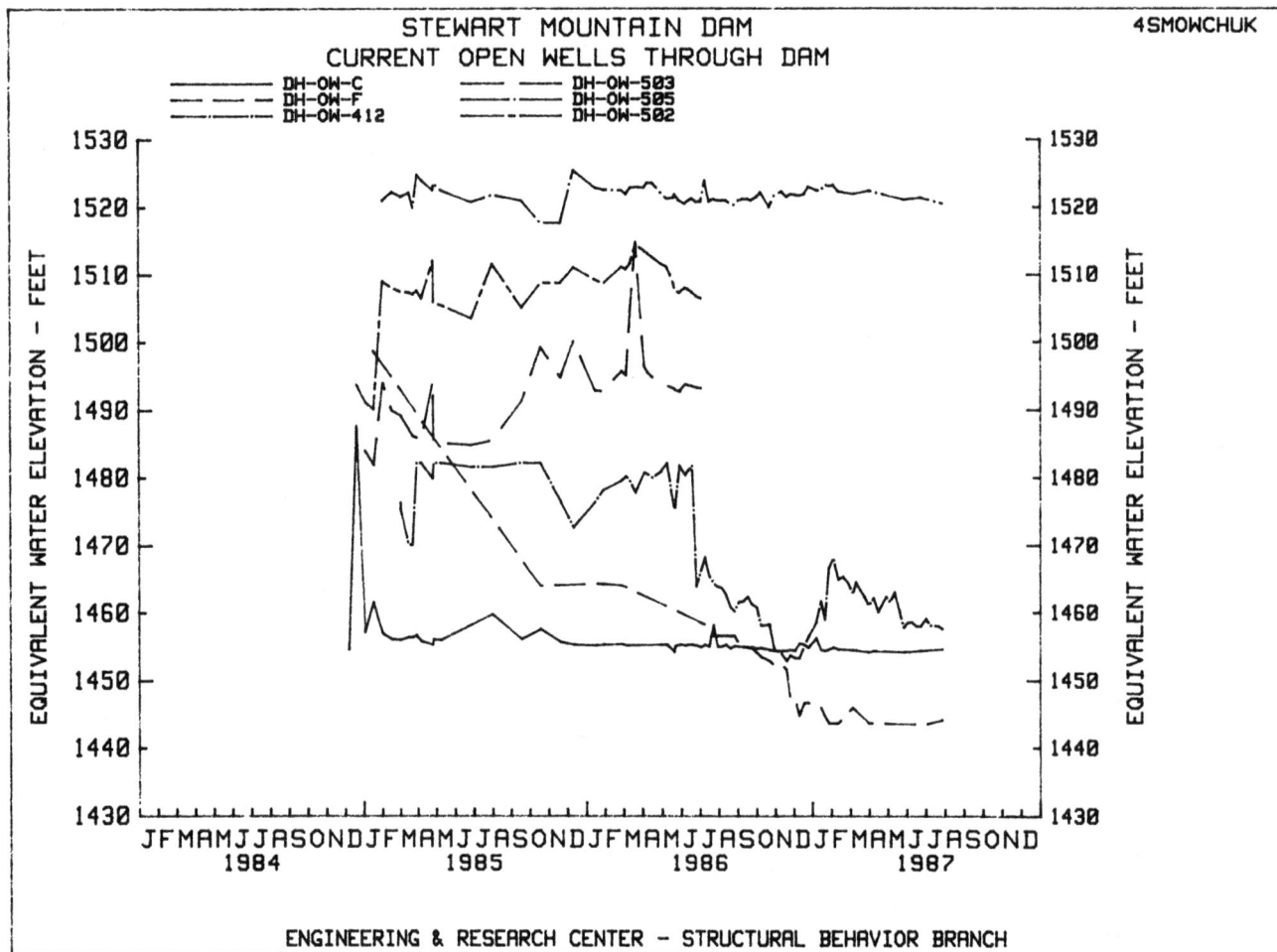

Figure 2-4.—Data plots on observation wells.

The air will also decrease the permeability of the porous tube. Thus, changes in the volume of air that may accompany pore pressure fluctuations tend to retard the flow of water into the standpipe and slow the piezometer's response. This condition rarely occurs in piezometer installations at concrete dams and can generally be prevented by ensuring that the porous tube and sand backfill are thoroughly saturated during installation.

To isolate hydrostatic pressures in a relatively thin aquifer, it may be necessary to limit the length of the porous space, thereby sacrificing some sensitivity. This factor can be partially offset by using a hole or influence zone of larger diameter in that zone.

   c. *Description of Devices.*—Porous-tube piezometers consist of a porous-tube assembly, standpipe, backfill material, protective casing, water-level indicator and, in some installations, a protective cover pipe. Figure 2-6 illustrates a typical porous-tube installation. A conventional porous-tube installation contains a single standpipe; however, a two-tube system has been developed that provides additional access if one standpipe should become plugged.

   (1) *Porous-Tube Assembly.*—Porous-tube assemblies may consist of alundum, carborundum, or high-density polyethylene porous tubes, a reducer bushing, and a pipe plug. Figure 2-7 shows typical construction details. The porous tube is usually cemented to the plastic standpipe with a liquid-weld cement or epoxy.

   (2) *Standpipe.*—The standpipe from the porous-tube assembly to the readout level is usually a ½- or ¾-inch (13- or 19-mm) diameter PVC pipe furnished in straight sections. Past experience using a ⅜-inch (10-mm) diameter pipe furnished on rolls has indicated problems with unrolling in cool weather and the tendency to kink, which may cause the water-level probe to hang up during monitoring.

   Threaded, internal couplings are recommended because of their ease of installation and the problems

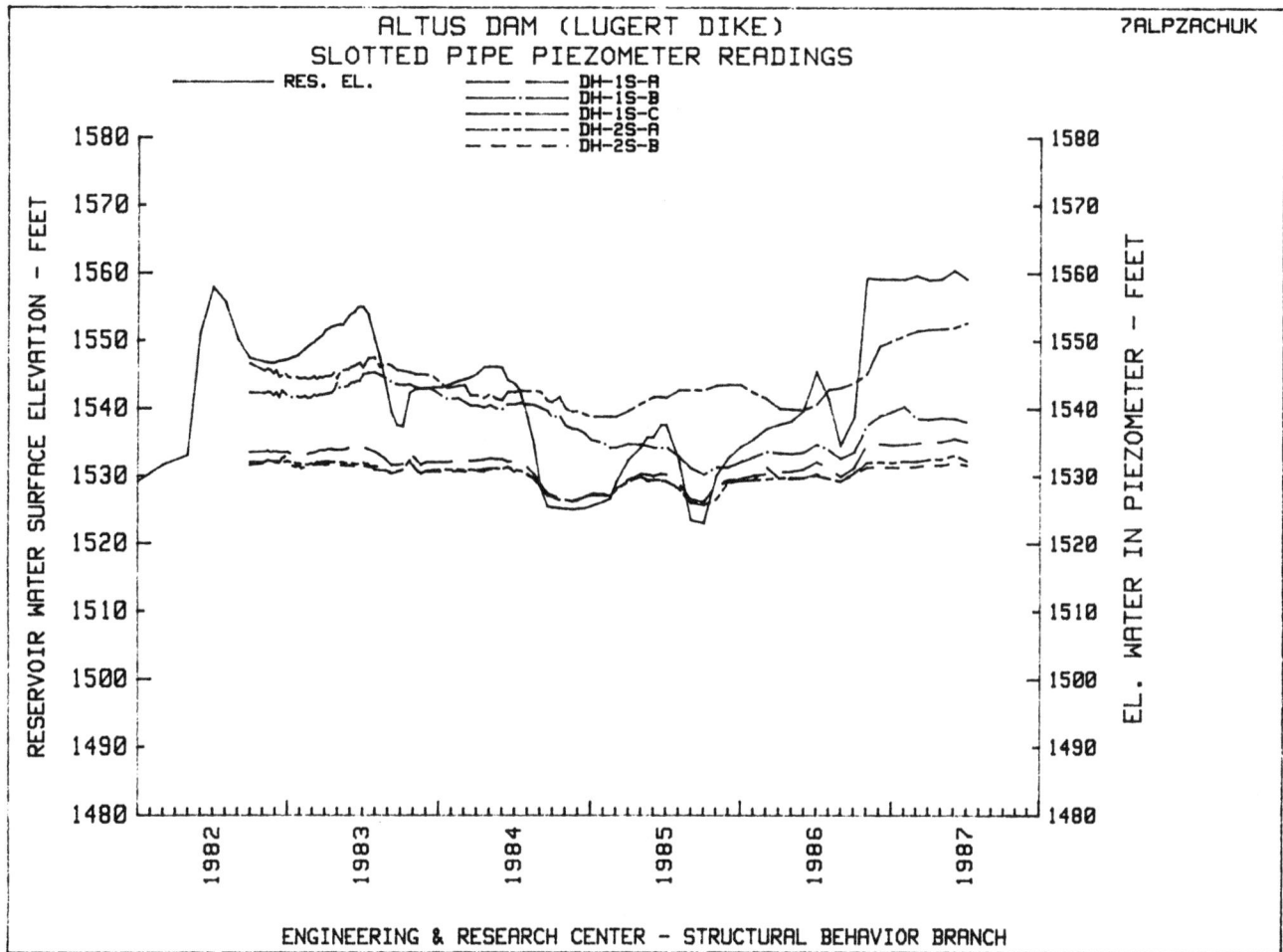

ALTUS DAM (LUGERT DIKE)   ?ALPZACHUK
SLOTTED PIPE PIEZOMETER READINGS

Figure 2-5.—Data plots on slotted-pipe piezometers.

encountered using liquid-weld cement on slip couplings in cold weather. The slip couplings have occasionally been found to be machined poorly, and the cement could cause internal pipe obstructions. Also, external couplings may create an unnecessary obstruction in the hole.

(3) *Drill Hole Requirements.*—A minimum drill hole diameter of 4 inches (102 mm) is required for the installation of a single porous-tube piezometer. The 4-inch space allows sufficient room for an adequate sand backfill around the tube. If two piezometers are to be installed in the same hole but at different elevations, a 6-inch (152-mm) diameter drill hole should be used.

(4) *Backfill Material.*—Graded sand, concrete sand, bentonite pellets, and/or bentonite-sand mixes are usually used for backfill. Sand is placed around the porous tube and bentonite is used to seal the hole above and below the porous tube elevation.

(5) *Water-Level Indicator.*—The water-level indicator used to monitor the installation is the same as the one described in section 2.2 c. (4).

(6) *Protective Casing.*—Some installations require a protective casing around the piezometer tip and the standpipe. If used, the protective casing is normally constructed of 3- or 4-inch (76- or 102-mm) diameter PVC pipe. The casing is perforated or slotted in the area of the influence zone to be measured, and the perforations are covered with metal screening or filter fabric to prevent sand backfill intrusion.

(7) *Protective Cover Pipe.*—A protective cover pipe, if required, is identical to the one described in section 2.2 c. (5).

d. Installation Techniques.—The installation of a porous-tube piezometer is the same as described for an observation well in section 2.2 d., except that bentonite seals are required above and below the porous-tube inlet (sec. 2.3).

23

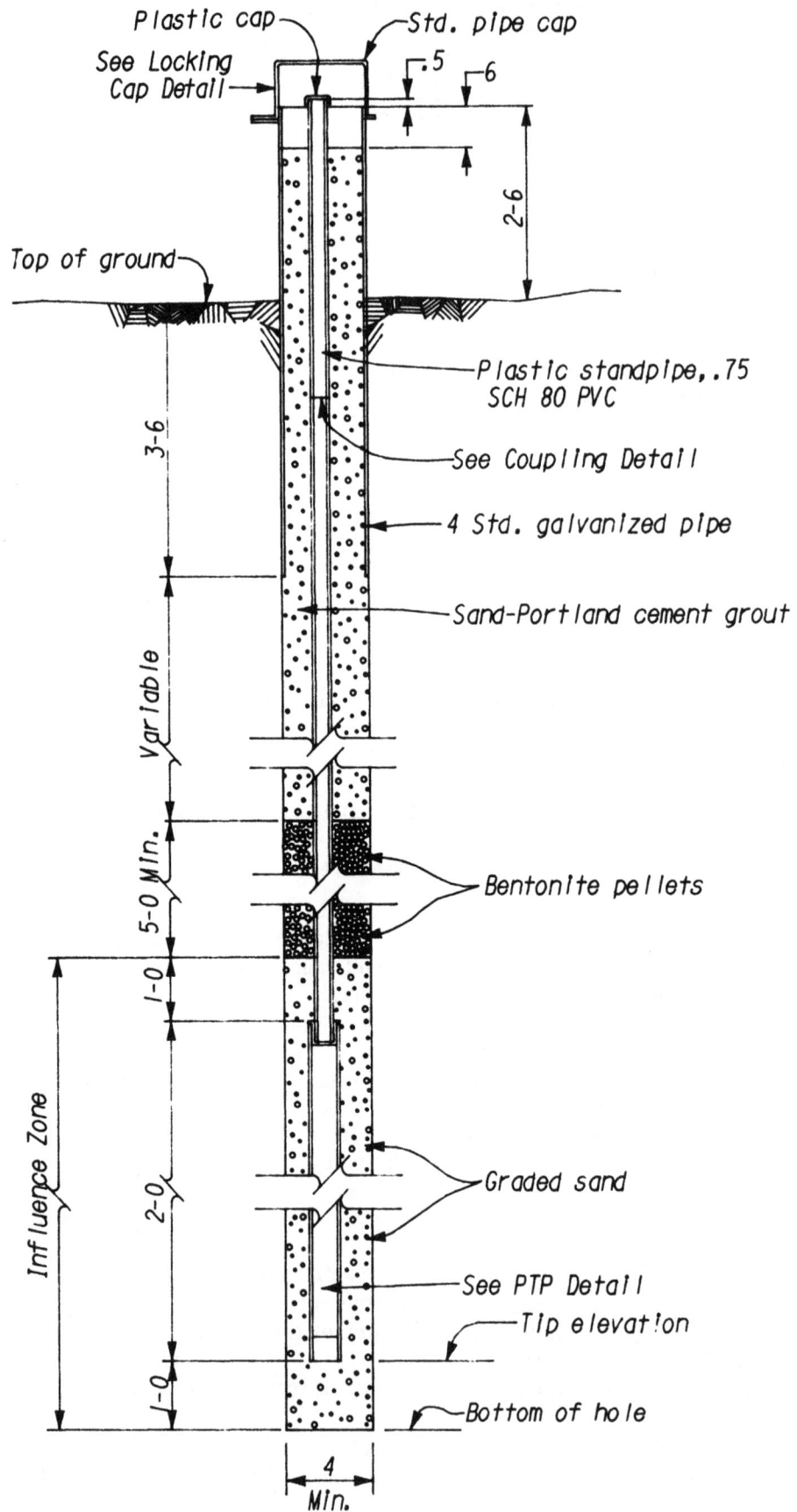

Figure 2-6.—Typical porous-tube piezometer installation. From 963–D–126.

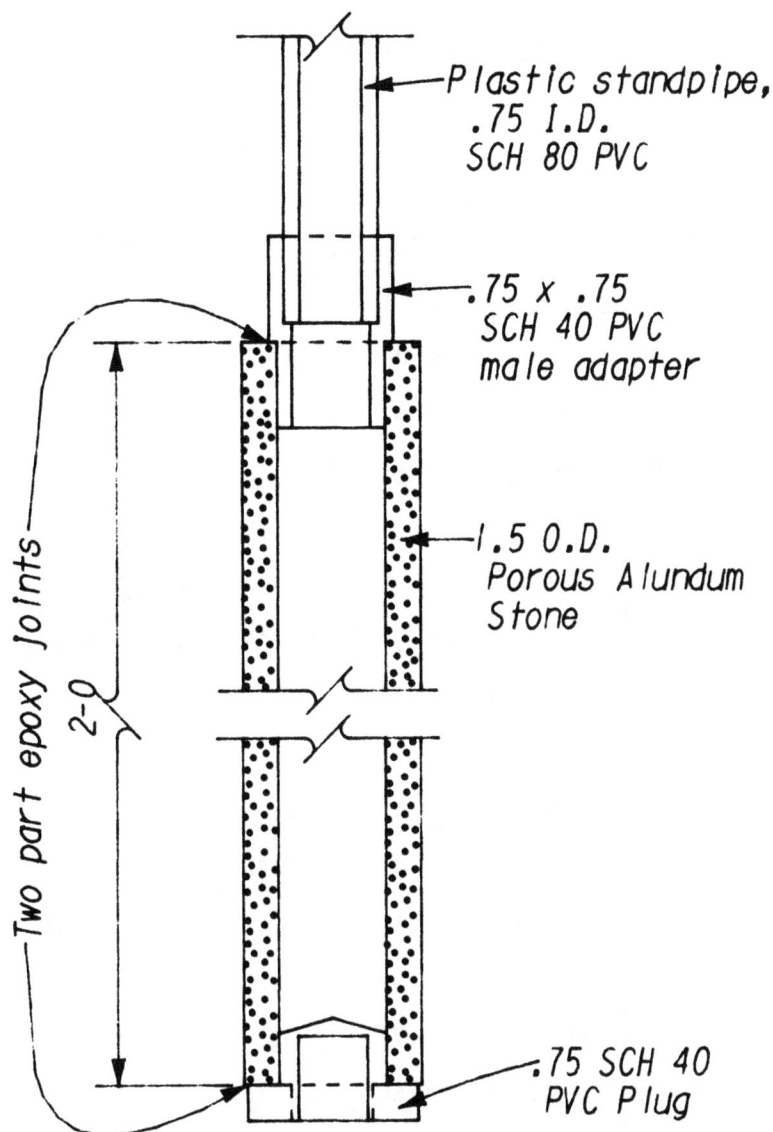

Figure 2-7.—Details of a porous-tube piezomter assembly. From 963-D-126.

e. *Monitoring Procedures.*—The monitoring procedures used for porous-tube piezometers are the same as those described in section 2.2 e. for observation wells. Data are entered on a field data form as shown of figure 2-8.

f. *Data Acquisition and Processing.*—Readings are obtained in feet or meters of depth to the nearest 0.01 foot (0.003m) from top of standpipe. The depth to the water level is subtracted from the elevation at top of standpipe, which yields the water-level elevation.

The data are then transmitted to the Denver Office where the data are processed, reviewed, and action taken as indicated in section 1.25 and on figure 1-1. Example data plots are shown on figure 2-9.

B. Closed System Piezometers

2.5. Vibrating-Wire Piezometers.—a. *Usage.*—Vibrating-wire piezometers are used effectively in both foundation and abutment applications to monitor pore water pressures. In some installations, vibrating-wire piezometers have been used to check the accuracy of other adjacent instruments; they may also be used in situations where negative pore pressures may be encountered.

b. *Advantages and Limitations.*—The advantages of using vibrating-wire piezometers include a relatively easy readout, simple maintenance, a relatively short lag time in readings, ability to monitor negative pore

25

POROUS-TUBE
PIEZOMETER READINGS

TO: DIRECTOR OF DESIGN & CONST.
ENGRG. & RESEARCH CENTER
P.O. BOX 25007, DEN. FED. CNTR.
DENVER, COLO. 80225-0007
ATTENTION: 3352

MODIFIED................FOR

Dam..... Brantley

Project..... Brantley

Ref. Dwg..... 963-D-126 .....Reservoir Water El.(1) .....--.....

Date of Observations..... 05/08/87

Observer..... JAJ .....Sheet.....2.....of.....5

Tailwater El.(1) .....

POROUS TUBE NO.	LOCATION (2)		ORIGINAL (3) EL. OF POROUS TUBE	EL.-TOP RISER TUBE (4)		SETTLE-MENT-TOP RISER TUBE	DISTANCE-TOP RISER TUBE TO WATER SURFACE	EL. WATER IN PIEZOMETER
	STATION	OFFSET		ORIGINAL	PRESENT			
						*		
PTP-1	36+48	100.3D/S	3260.15	3274.40	3274.40	--	DRY	--
PTP-2	39+48	90.5D/S	3262.11	3275.91	3275.91	--	DRY	--
PTP-3	54+00	155.4D/S	3225.16	3242.66	3242.66	--	DRY	--
PTP-4	63+00	170.0D/S	3225.17	3242.17	3242.17	--	DRY	--
PTP-5	73+25	29.8U/S	3199.40	3242.95	3242.95	--	26.10	3216.85
PTP-6	73+25	30.2D/S	3195.95	3241.55	3241.55	--	26.25	3215.30
PTP-7	NOT INSTALLED							
PTP-8	NOT INSTALLED							
PTP-9	87+07	164.7D/S	3224.07	3238.47	3238.47	--	DRY	--
PTP-10	94+52	209.8D/S	3189.11	3223.46	3223.46	--	28.10	3195.36
PTP-11	126+00	30.0U/S	3196.30	3287.30	3287.30	--	81.60	3205.70
PTP-12	126+00	30.0D/S	3198.50	3284.50	3284.50	--	77.55	3206.95
PTP-13	126+00	30.0U/S	3250.11	3284.71	3284.71	--	DRY	--
PTP-14	126+00	30.0D/S	3250.16	3284.60	3284.60	--	DRY	--
PTP-15	130+01	139.8D/S	3218.06	3249.11	3249.11		NO ACCESS	
PTP-16	142+00	152.4D/S	3227.73	3244.73	3244.73	--	DRY	--
PTP-17	191+00	100.4D/S	3216.88	3271.73	3271.73	--	DRY	--
			U/S RIVER ELEVATION 3218.0					

(1) Record if appropriate. (2) Record distance U/S or D/S from ℄ Crest of dam, or location by coordinates. (3) Taken as bottom of porous tube. (4) Record all elevations and distances to 0.01 foot. *Use minus (–) to indicate heave.

Figure 2-8.—Field data form for porous-tube piezometers.

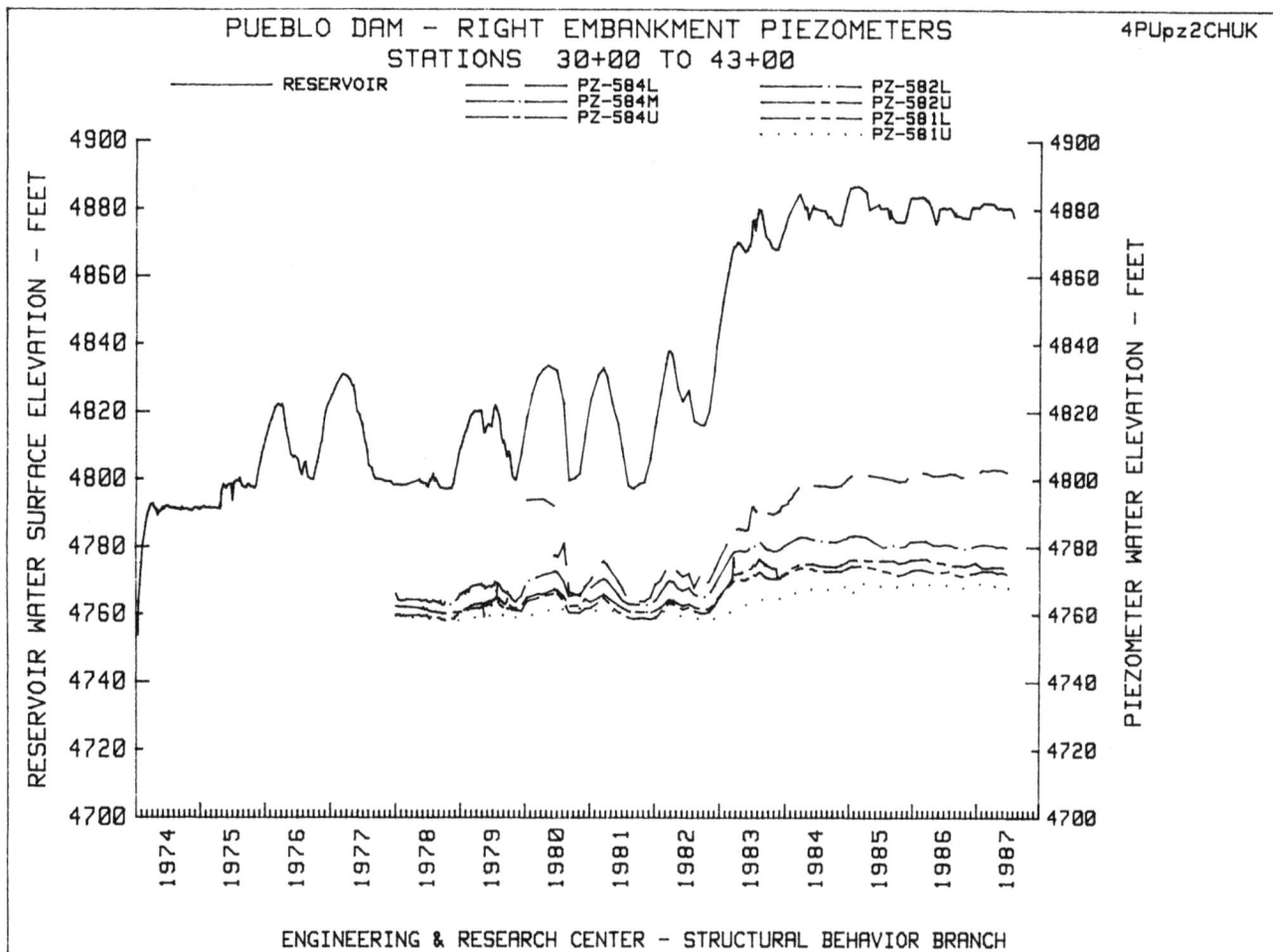

Figure 2-9.—Data plots on porous-tube piezometers.

pressures, elevations of cables and location of readout device are independent of location of piezometer tip, a central observation system may be used, and there are no problems with freezing.

Limitations include the inability to de-air the tip, thereby rendering long-term measurement of negative pore pressures or measurements in gaseous organic deposits somewhat questionable. In applications where very small variations in pressures are considered to be significant, corrections must be made for changes in barometric pressure and temperature, although this is not usually a significant problem on most dams.

c. Description of Equipment.—The piezometer tip contains a porous disk that allows water pressure to enter and press against a stainless steel diaphragm. A high-strength steel wire is fixed to the center of the diaphragm at one end and to a block at the other end. The wire is hermetically sealed within a stainless steel member and set to a predetermined tension during manufacture. Pressure applied to the diaphragm causes it to deflect, thereby changing the wire tension and resonant frequency of the wire. A coil/magnet assembly is used in conjunction with a readout device to "pluck" or vibrate the wire and to measure the wire's vibration frequency. The pore pressure may be calculated from the displayed readings using calibration charts, an equation, or by using a simple gauge factor.

d. Parts.—Vibrating-wire piezometer units consist of a piezometer tip, backfill materials, electrical cable, and a readout unit.

(1) *Piezometer Tip.*—As previously described, the piezometer tip assembly consists of a stainless steel body, porous disk, high-strength steel wire, stainless steel diaphragm, transducer for transmitting resonant frequency of wire, and a coil/magnet that causes wire to vibrate. Figure 2-10 shows details of this assembly. Such devices are available in pressure ranges from 25 to 5,000 lb/in^2 (172 to 34 474 kPa). The 250-lb/in^2 (1724-kPa) device is normally used for Bureau dams.

27

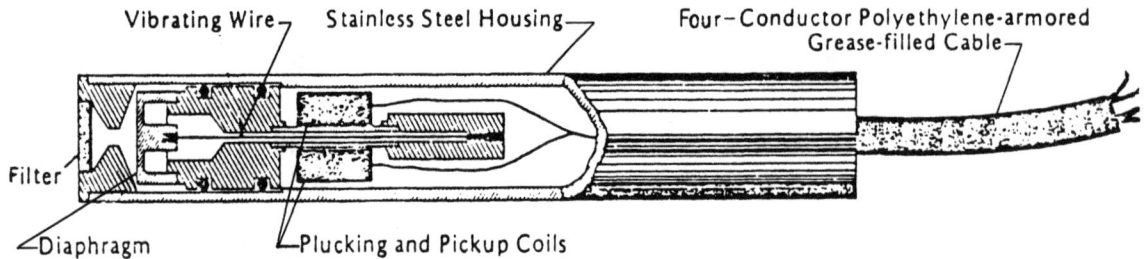

Figure 2-10.—Details of a vibrating-wire piezometer tip assembly.

(2) *Backfill Material.*—Typical backfill material in vibrating-wire piezometer installations consists of graded sand around the tip and bentonite pellets placed above and below the tip to isolate the tip area. Bentonite-sand mixtures are commonly used around portions of the electrical cables to prevent migration of water along the cables.

(3) *Electrical Cable.*—Each vibrating-wire piezometer is orderd with sufficient electrical cable to allow for the required length to reach from the piezometer tip to the readout location, plus an additional 10 feet (3.0 m) for each piezometer and an additional 10 percent for slack. Four conductor cables shielded with a neoprene jacket are used. The outside of the cable should be permanently marked by the manufacturer, every 10 feet along the length of the cable with the piezometer identification number to which the cable is attached.

(4) *Vibrating-Wire Readout Unit.*—Readout units must be capable of accepting an electrical signal from a vibrating-wire piezometer and displaying the frequency reading or converted engineering units on a digital display. A battery charger unit is also required, as is a supply of short electrical leads to connect electrical cables to readout unit.

e. *Piezometer Installation.*—(1) *Preinstallation Procedures.*—(a) Hole drilling and cleanout of the hole is performed in all piezometer installations.

(b) The sintered stainless steel filter on the end of each piezometer must be purged of air before installation. This is accomplished by removing the filter from piezometer tip and placing the filter in a bath of boiling water. The O-ring must first be removed from the filter housing.

(c) Piezometer is then reassembled while submerged in cool water. It is customary to do this by pouring a relatively small amount of the hot water used to boil the filter into a large container of cold water. The size of the container should be large enough to allow reassembly of the piezometer, including enclosing it in a sand-filled bag as described later, while keeping the peizometer submerged at all times.

(d) Place a cloth bag in the large container and partially fill the bag with graded sand. Place the piezometer in the center of the bag and pack the sand around the piezometer. The top of the bag is then tied when full and the entire bag wrapped with wire mesh screen. The entire unit should be kept submerged until installation.

(e) Using the readout unit, obtain an initial reading on the piezometer while it is still in the container of water. This reading is the zero or initial reading for that particular piezometer. Because all piezometers are individually calibrated, it is very important to note and record which piezometer is installed at what location. The individual transducer numbers must therefore be accurately recorded, preferably in more than one location.

(2) *Installation in Drill Holes.*—Upon completion and cleanout of the drill hole to a depth of at least 1 foot (0.3 m) below the desired piezometer elevation, compacted, saturated, graded sand is placed in the hole up to the piezometer elevation. The piezometer is then lowered to the desired elevation and a reading taken to ensure that it is operating properly. If piezometer is operating correctly, the installation continues with the rest of the influence zone being backfilled with the saturated, compacted, graded sand. Then, a bentonite pellet seal is placed and tamped. The entire process is repeated if any higher level piezometers are to be installed in the same hole. Following compaction of upper bentonite seal, all slack in the electrical cables is removed and remainder of hole is backfilled with a sand-bentonite mixture. If a casing was used, the casing is pulled as filling progresses or after backfilling is complete.

f. *Monitoring Procedures.*—Vibrating-wire piezometers are supplied with a reference reading that corresponds to the period (inverse frequency) of the initial wire tension that was set at the factory. The barometric pressure is also noted. Upon arrival at the project site, the initial gauge readings should be checked. Colored lead wires usually are for the pressure transducer and, when included, white wires connect to a temperature

28

reading thermistor. For example, to check the initial reference reading on IRAD-type gauges, set readout unit gauge select switch to "Gauge Type 3," mode switch (where present) to "Normal," connect leads to piezometer cable wires, and note the reading. The displayed reading should be within 20 digits of the reading supplied with the gauge. The exact difference in readings will depend upon the temperature and barometric pressure at the test site and scale range of the readout device.

The peizometer tips are sealed during manufacture; therefore, corrections for barometric pressure differences must be accomplished by determining the difference between the site barometric pressure and the barometric pressure noted at the time of factory calibration. This difference is then added to the calculated pressure at the site.

The basic relationship between the deflection of the diaphragm in a piezometer and the vibration frequency or period of vibration of a vibrating wire attached to the diaphragm is:

$$\text{Deflection} = (\text{constant})(\text{frequency})^2 = \frac{(\text{constant})}{(\text{period})^2}$$

where the constant takes into consideration the diameter and length of the wire. Because pressure is proportional to diaphragm deflection, the relationship between the applied pressure P and the readout unit is expressed by the equation:

$$P = G(L_0 - L_1) \tag{1}$$

where:

P = pressure in pounds per square inch (kilopascals),
G = gauge constant (factor) supplied with piezometer, constant is adjusted depending upon whether inch-pound or SI-metric units are used;
L_0 = initial zero reading under zero pressure at site during installation, and
L_1 = reading under actual pressure at site.

Some readout devices can be read directly with the gauge switch set in "Linearized" mode. In this case, the initial zero reading has been recorded at time of piezometer installation and a new L_1 reading is all that is required to use equation (1). Alternatively, for approximate pressure values, a calibration graph similar to the one shown on figure 2-11 may be used directly by scaling off the graph. With gauge switch in "Normal" mode, the reading displayed must be converted using the conversion tables supplied with the unit. Similarly, temperature corrections may be made using a temperature constant supplied with the unit as follows:

$$PT_1 = G(L_0 - L_1) - K(T_0 - T_1) \tag{2}$$

where:

K = a supplied temperature constant that should be adjusted depending upon whether inch-pound or SI-metric units are used;
T_0 = initial temperature at calibration (supplied by manufacturer), in °F or °C;
T_1 = current temperature of piezometer at site, in °F or °C; and
$P, G, L_0,$ and L_1 = as previously defined in equation (1)

For example, specific instructions for a gauge reading using an IRAD-type gauge (MB-6-LU) are as follows:

1. Plug jumper leads into top panel of readout device.
2. Attach clips on jumper leads to colored leads of piezometer cable (sequence is not important). The white leads are the thermistor leads.
3. Set gauge type for "Type 3" gauge.
4. Set auto/manual switch on "auto."
5. Set normal/lineral switch on "linear."
6. Turn power on.
7. If a colon appears in display, unit needs to be charged. Charging should not exceed a 16-hour period.
8. Reading will be displayed as a four-digit number, which should be recorded without decimals. This is the N_0 (initial) reading. It is not necessary to convert the reading to linearized form, this will be done at the E&R Center. A temperature reading is then taken and recorded.
9. Disconnect all leads.
10. Turn off power. Unit will automatically turn off in about 4 minutes if unattended.

PIEZOMETER CALIBRATION

Model No. _____4500S_____   Date _____10 March 83_____

Serial No. ____255T-100_____   Temp. ____64.4 °F_____

Gauge Factor, G ___0.0499_____   Barometer ____29.86_____

Temp. Coeff., C ___0.0025 (lb/in²)/ °F____   Zero Rdg., L_0 ____3273_____

                                         Period, _____5527_____

$$Pressure = (L_0 - L_1)\, G$$

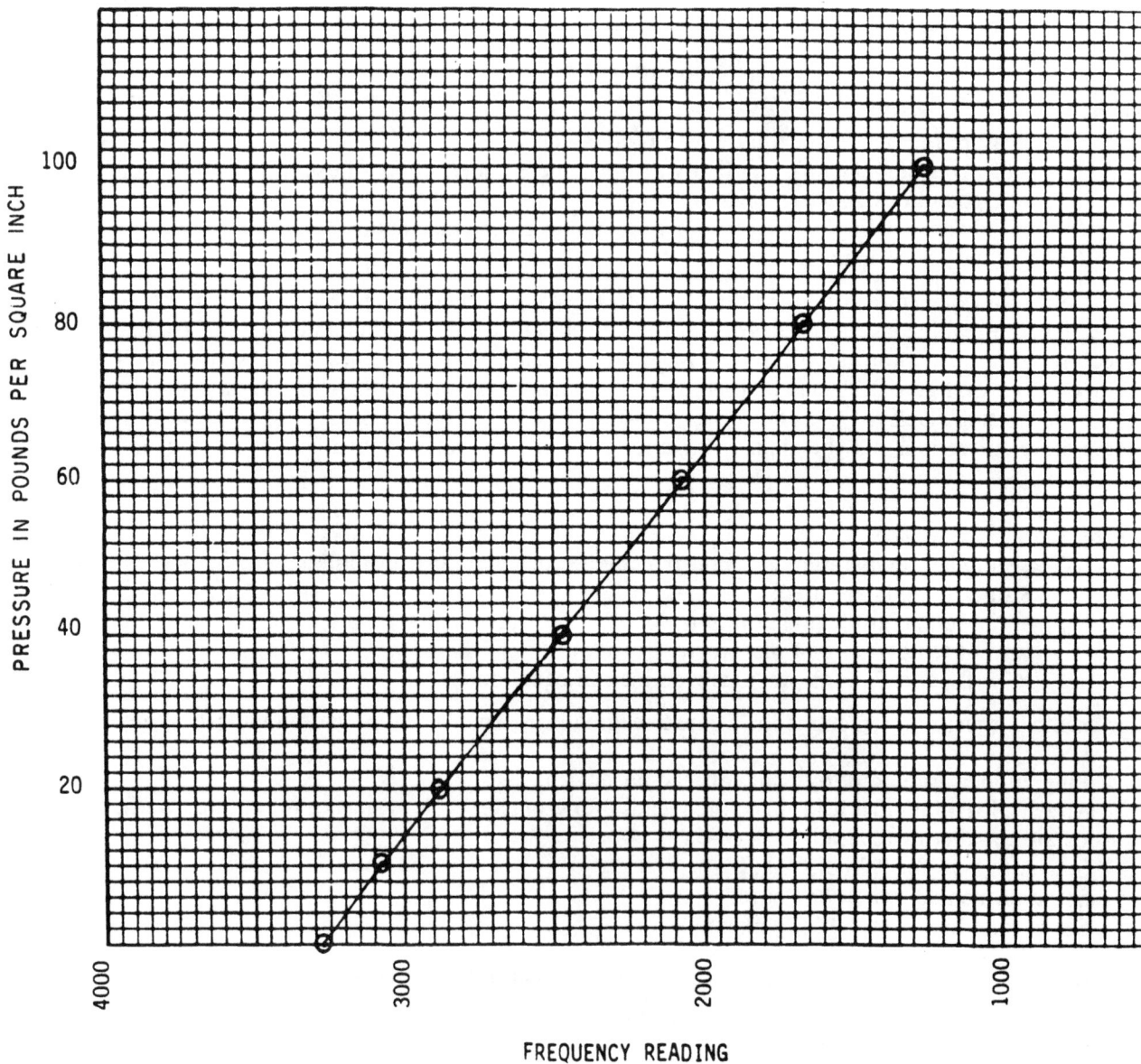

Figure 2-11.—Calibration graph for a vibrating-wire piezometer.

g. *Data Processing and Review.* If the pressure readings obtained from equation (1) are in pounds per square inch, readings are then converted to the piezometric elevation by multiplying by a factor of 2.31 to obtain pressure in feet of water, and then algebraically adding this value to the tip elevation to obtain the actual piezometric elevation. If the pressure readings are in kilopascals, multiply the reading by 0.102 to obtain the actual piezometric elevation in meters. Figure 2-12 shows a sample data sheet.

The data are transmitted by mail or computer to the Denver Office for analysis. Upon arrival, the data are processed, reviewed, and reports prepared as discussed in section 1.25 and shown on figure 1-1. Example data plots are shown on figure 2-13.

2.6 Uplift Pressure Measurements.—a. *Usage.*—It is important to determine the magnitude of any hydraulic pressures that may be present at the base of a dam due to percolation or seepage of water along underlying foundation seams or joint systems after the reservoir is filled. Measured values of uplift pressure may also indicate the effectiveness of foundation grouting and of the drainage systems contained within and below the dam, see figure 2-14.

b. *Methods.*—A system of piping is installed at several stations in the dam in the first lift just above the contact between the foundation rock and the base of the dam. Drill holes with installed drainpipes extend below the dam both upstream and downstream of the grout curtain; these pipes extend into one or more of the lower galleries in the dam. The pipes are usually fitted with a tee-section and a Bourdon-type pressure gauge for observing water pressures. A shutoff valve is usually included on one leg of the tee. Thus, the uplift pressure may be determined by either reading a pressure gauge or by sounding with a water measurement probe.

FORM D-3352-4

VIBRATING-WIRE PIEZOMETERS

TO: ASSISTANT COMMISSIONER-
ENGINEERING AND RESEARCH
ENGRG & RESEARCH CENTER
P.O. BOX 25007, DEN. FED. CNTR.
DENVER, CO 80225
ATTENTION D-3352

Dam_____   Observer(s)_____

Reservoir Elevation_____   Date_____

Tailwater Elevation_____   Sheet_____of_____

PIEZOMETER NUMBER	SERIAL NUMBER	GAUGE FACTOR	INITIAL READING (FREQ)	CURRENT READING (FREQ)	PIEZOMETRIC HEAD (FT)	PIEZOMETER TIP ELEVATION (FT)	PIEZOMETRIC ELEVATION (FT)	TEMPERATURE CORRECTION		PIEZ. ELEV. AFTER CORRECTION (FT)
								RESISTANCE (OHMS x 10³)	TEMPERATURE (°F)	
		G	F_0	F_1	H*	A	H+A			

* $H = 2.308 [G (F_0 - F_1)]$

Figure 2-12.—Field data form for vibrating-wire piezometers.

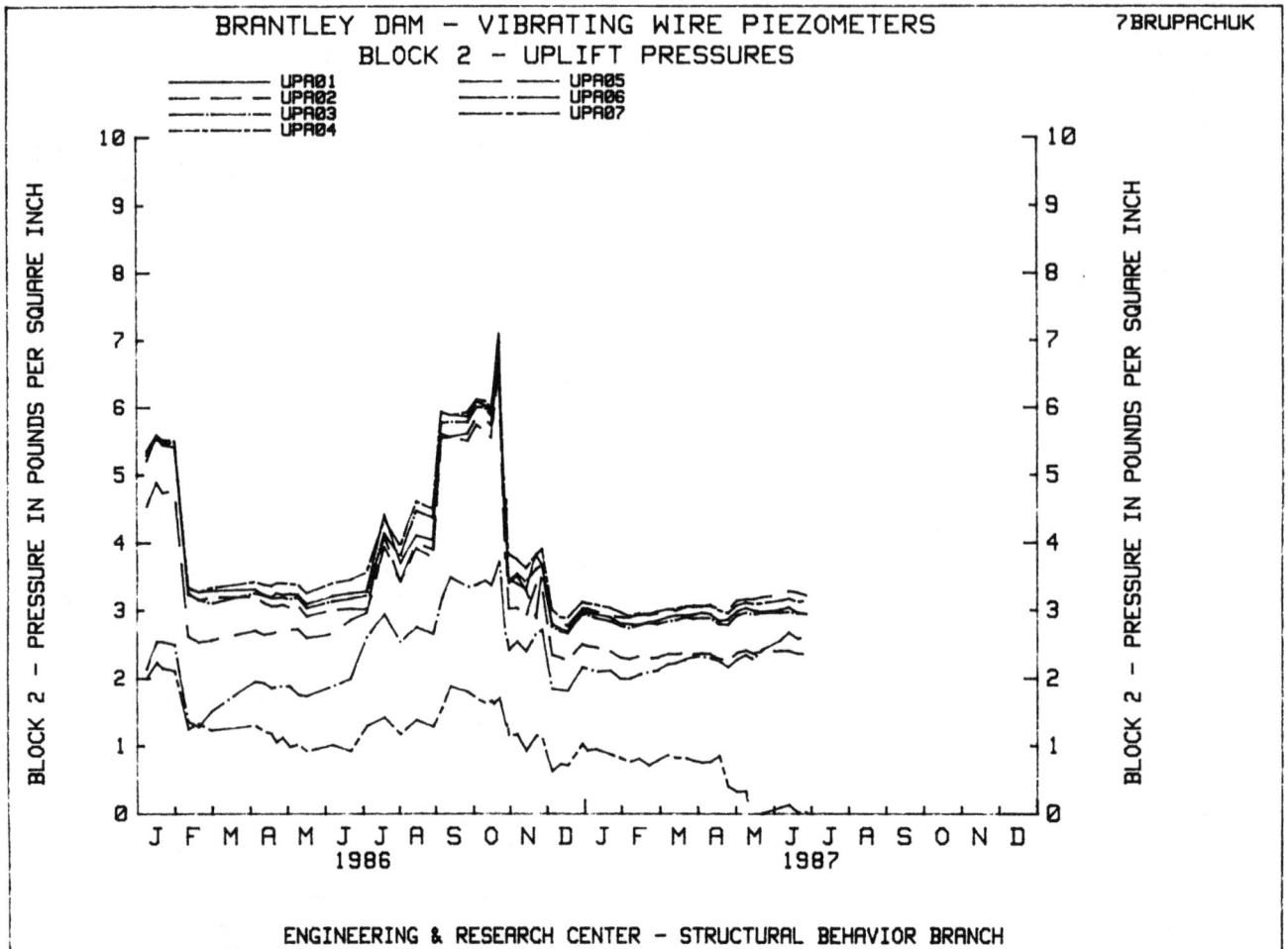

Figure 2-13.—Data plots on vibrating-wire piezometers. (Sheet 1 of 2).

Generally, pipes placed upstream of the grout curtain are read with a pressure gauge and those placed downstream of the grout curtain are read by sounding or with a pressure gauge. Although readings may be taken at any time, it is common practice to leave the shut-off valve open and to close the system the day before readings are to be taken. Then, after readings are taken, the valves are reopened. Therefore, in addition to the existing uplift pressures, readings taken in this manner indicate the uplift pressures that may be achieved if one or more systems of drains should become inoperative through blockage. Figure 2-15 shows a typical installation, and figure 2-16 shows a Bourdon-gauge installation.

Uplift pressure measurements are recorded either in feet (meters) of water or in pounds per square inch (kilopascals) pressure, and forwarded to the Structural Behavior Section for analysis and processing as discussed in section 1.25. Data plots are then prepared as indicated on figure 2-17.

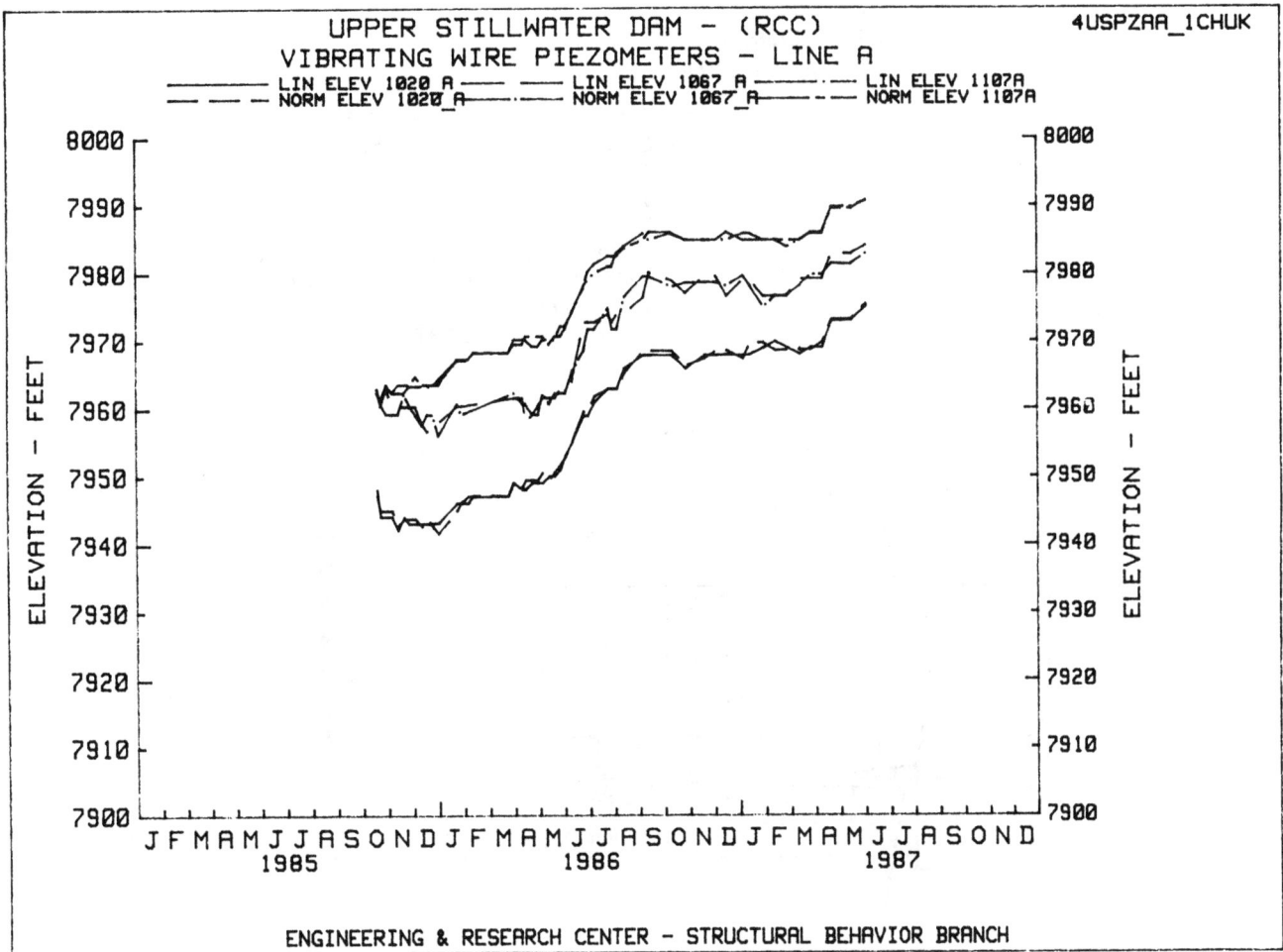

Figure 2-13.—Data plots on vibrating-wire piezometers. (Sheet 2 of 2).

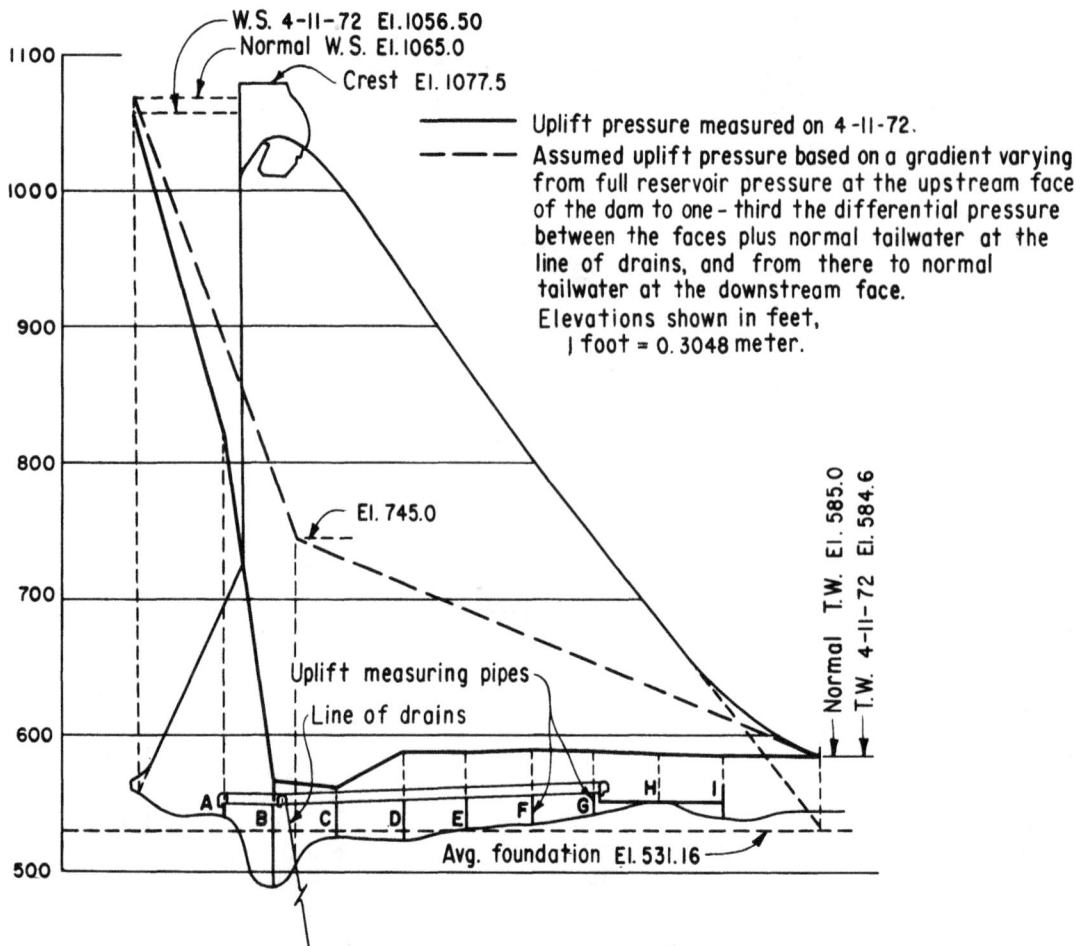

Figure 2-14.—Uplift pressure gradient through a dam. 288-D-2959.

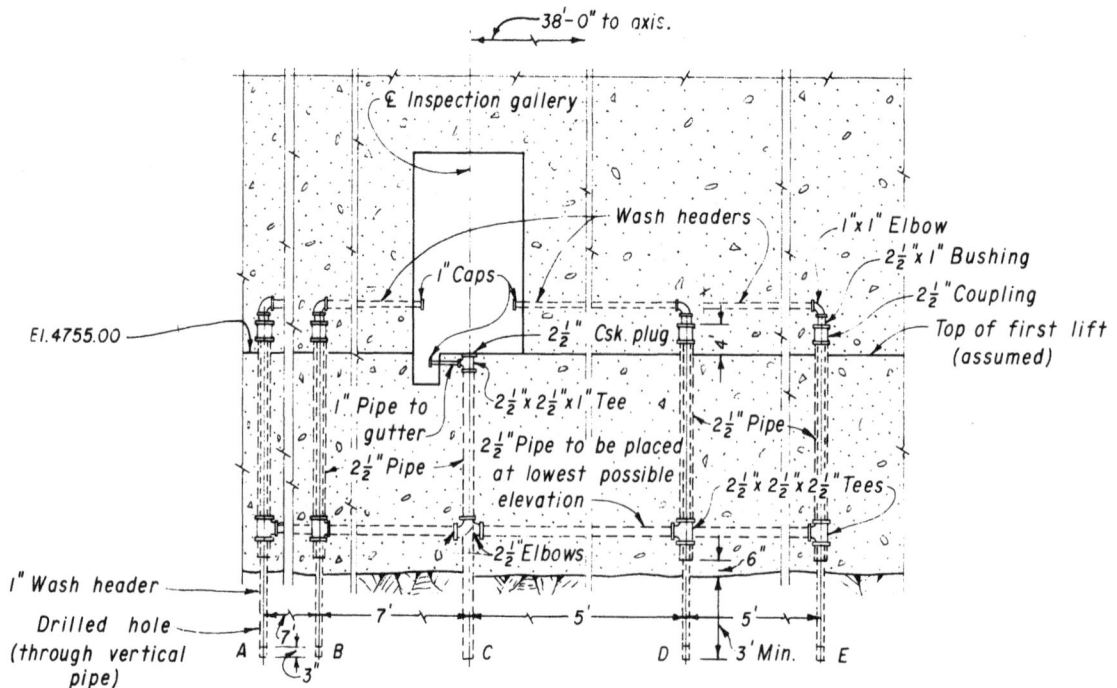

Figure 2-15.—Typical installation of an uplift pressure measuring system. From 382-D-1350.

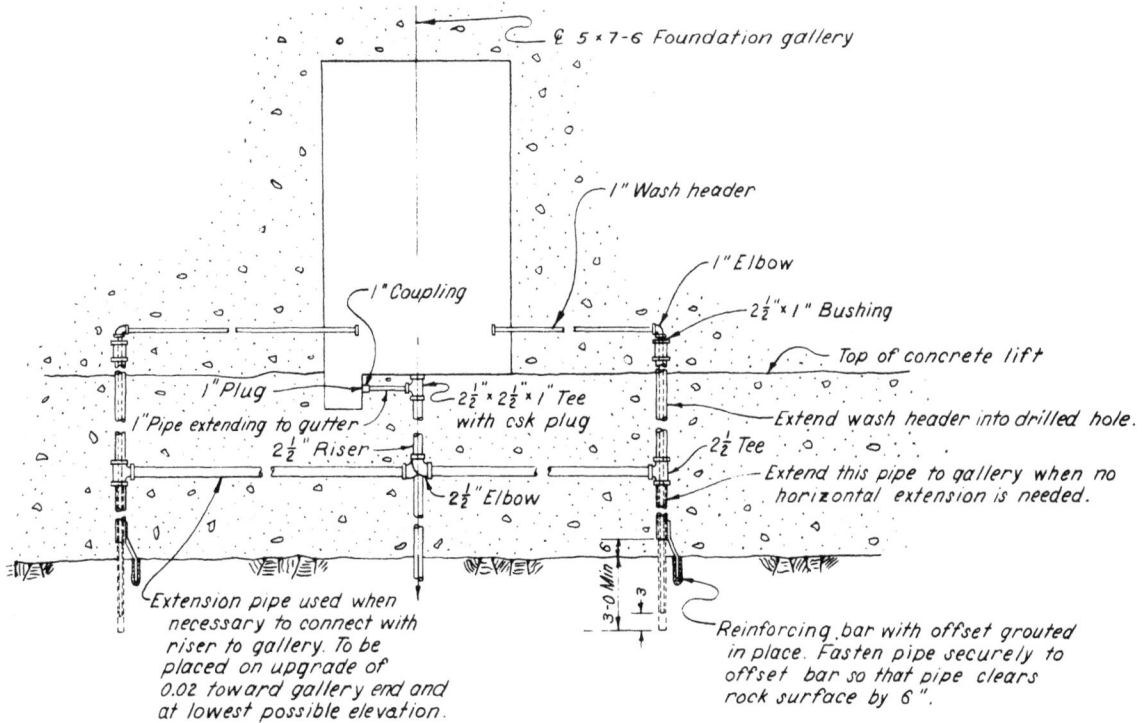

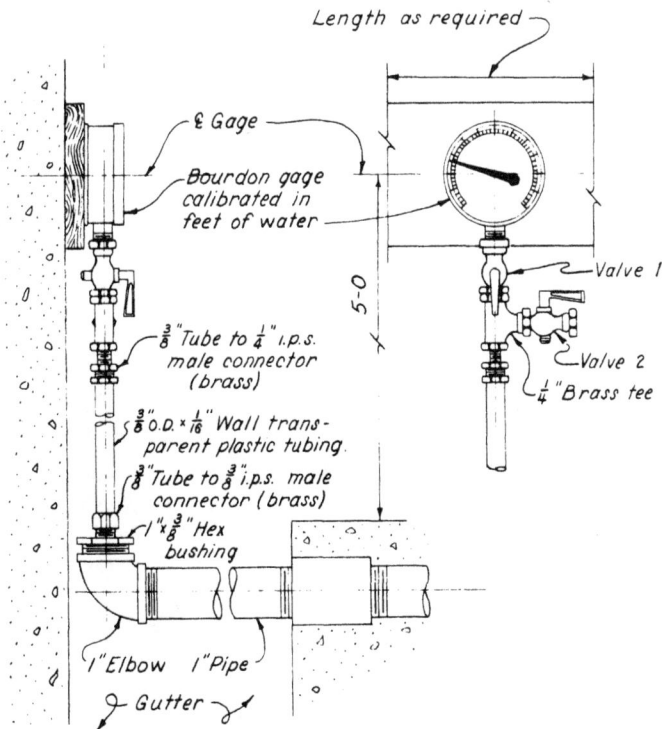

Figure 2-16.—Typical Bourdon gauge installation. 465–D–765.

35

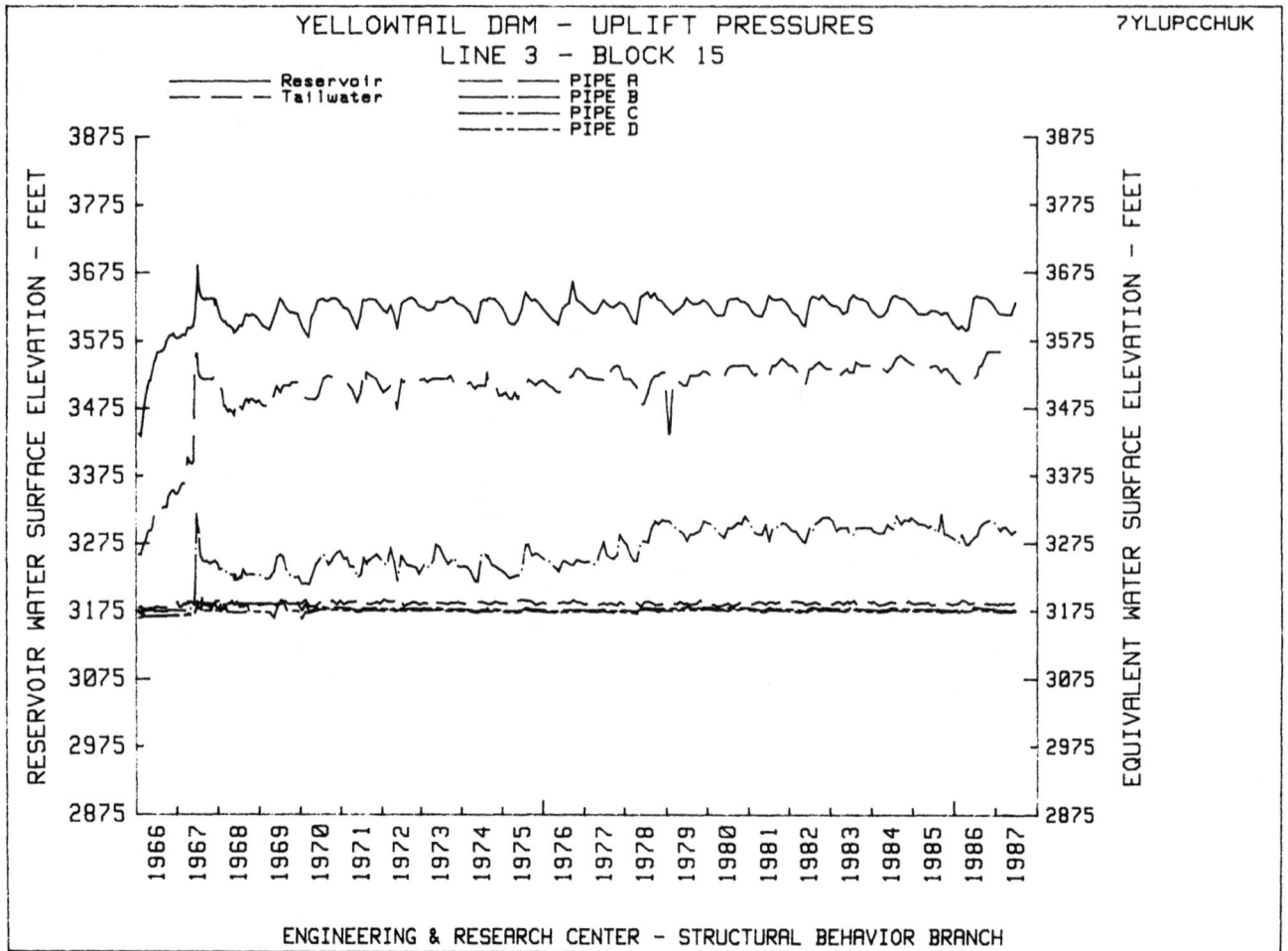

Figure 2-17.—Data plots on uplift pressures.

Chapter 3

SEEPAGE MEASUREMENT DEVICES

A. Currently Used Devices

3.1 General.—Seepage measurement devices are installed at dams to measure the amounts of seepage through, around, or under dams. Drain outlets are commonly used as seepage measurement points. The seepage water should be tested to determine its chemical composition because chemical changes may indicate progressive dissolution or erosion in a dam foundation or abutment.

Currently used seepage measurement devices described herein include various types of weirs, calibrated catch containers, and flowmeters. Detailed information concerning other types of weirs and flumes, including formulas and tables for computing flow quantities, is not covered in this manual; however, this information is covered in the Bureau's *Water Measurement Manual* [4]. Experimentation is also currently underway using a modified or "cutthroat" flume.

3.2 Weirs.—**a. *General.***—Their weir is one of the oldest, simplest, and most reliable types of devices used to measure the flow of water. If sufficient fall is present in the channel and the quantity of water to be measured is relatively small, the weir is a very serviceable and economical measuring device. For concrete dams, most internal seepage and/or drain flow is measured in the gallery drainage gutters, which are about 1-foot (0.30-m) square. The most common type of weir in use is the 90° V-notch weir. Other weir types such as rectangular or trapezoidal may be used for larger flow quantities. These types consist of overflow structures installed in a section of open channel.

b. *Standard 90° V-Notch Weir.*—The triangular or V-notch thin-plate weir is an accurate flow measuring device particularly suited for small flows. The weir usually used by the Bureau is the 90° V-notch shown on figure 3-1; however, 22.5° and 45° weirs are also used.

The crest of the standard 90° V-notch weir consists of a thin plate with the sides of the notch being inclined 45° from the vertical. This weir operates as a contracted weir and all conditions for accuracy stated for the standard contracted rectangular weir apply. The minimum distances from the sides of the notch to the channel sides should be at least twice the head on the weir. This distance should be measured from the intersection points of the maximum water surface with the ends of the weir. The minimum distance from the crest to the pool bottom should be measured from the point of the weir's notch to the channel floor.

Because the V-notch weir has no crest length, the head required for a small flow through the weir is greater than that required with other types of weirs. This is an advantage for small discharges because the nappe will spring free of the crest, whereas the nappe would cling to the crest of another type of weir and make the measurement meaningless. Although Cipolletti and rectangular weirs that have a crest length of about 6 inches (150 mm) are sometimes used for measuring small flows, they are not as accurate or as sensitive as the V-notch weir for such flows and are not recommended. The 22.5° and 45° weirs are more accurate for smaller flows than the 90° V-notch weirs.

c. *Selection of Weir Types.*—Each of the weirs used by the Bureau has characteristics that fit that weir for particular operating conditions. In general, for best accuracy, a V-notch weir or a rectangular suppressed weir should be used. Cipolletti weirs and contracted rectangular weirs are very useful for many applications; however, they both have end contractions and have not been investigated experimentally as thoroughly as the suppressed rectangular and V-notch weirs.

Usually, the range of flows to be measured by a weir can be fairly well estimated in advance. With this range in mind, the following points should be considered:

1. The 90° V-notch weir is the preferred type for measuring discharges less than 1 ft³/s (0.03 m³/s), and it is as accurate as other types for flows from 1 to 10 ft³/s (0.03 to 0.28 m³/s). Therefore, this weir is well suited for discharges up to and a little higher than 10 ft³/s if sufficient head is available. For flows less than 0.5 ft³/s (0.015 m³/s), the 22.5° or 45° V-notch weir is preferred.

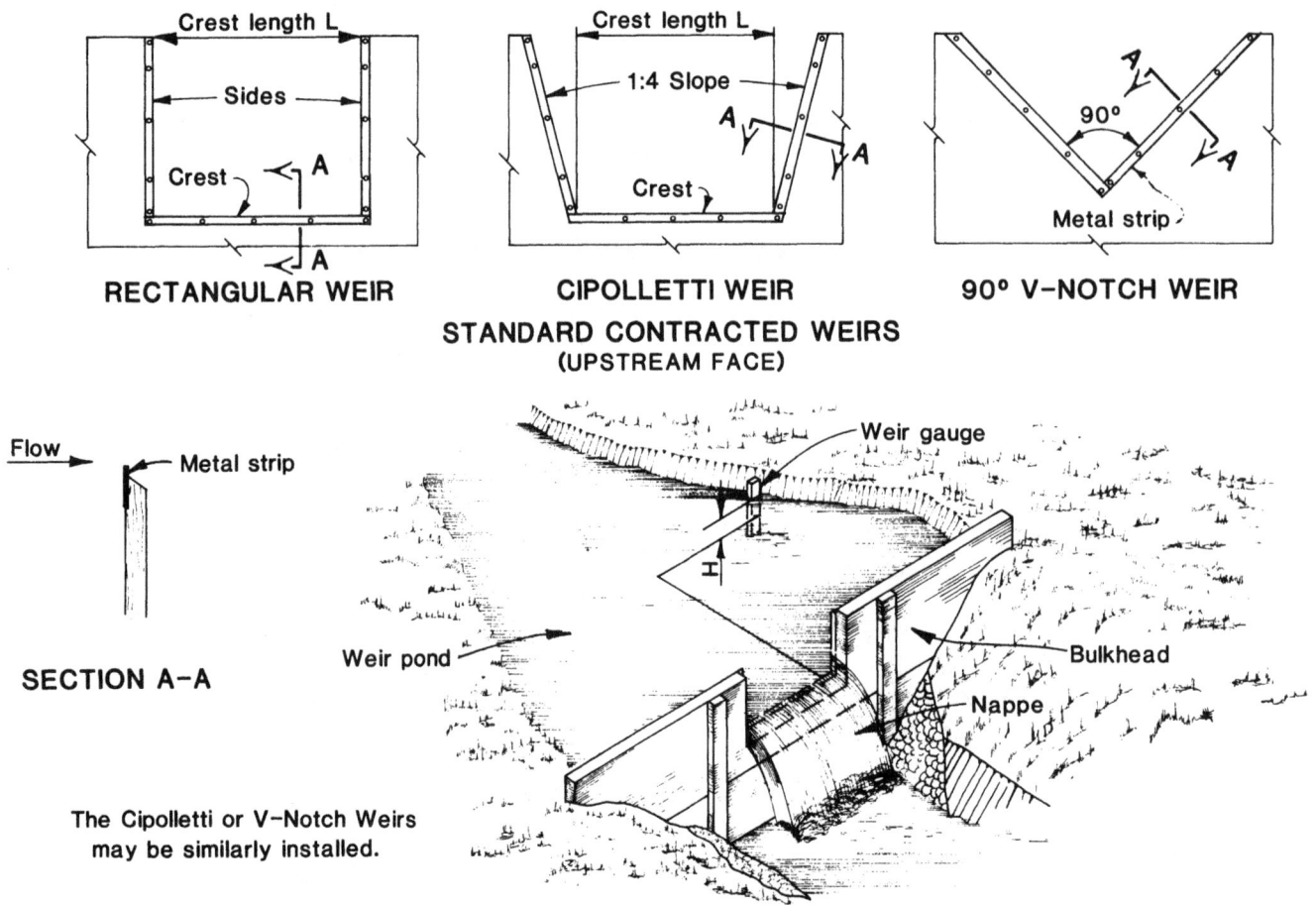

RECTANGULAR WEIR

CIPOLLETTI WEIR

90° V-NOTCH WEIR

STANDARD CONTRACTED WEIRS
(UPSTREAM FACE)

SECTION A-A

The Cipolletti or V-Notch Weirs
may be similarly installed.

Figure 3-1.—Standard contracted weirs, and temporary bulkhead with contracted rectangular weir discharging at free flow. 103-D-858.

2. The minimum head should be at least 0.2 foot (0.06 m) to prevent the nappe from clinging to the crest and because, at smaller depths, it is difficult to get sufficiently accurate gauge readings to calculate reliable flow quantities.

3. The length of rectangular and Cipolletti weirs should be at least three times the head.

4. If possible, the crests should be placed high enough so that the water flowing over them will fall freely, leaving an airspace under and around the jets. If submergence is permitted, special computations and reduced flow measuring accuracy may be expected.

   d. *Discharge Measurements.*—The rate of flow or discharge in cubic feet per second or cubic meters per second over the crest of a standard contracted rectangular weir or standard Cipolletti weir is determined by the head H in feet or meters and by the crest length L in feet or meters. The discharge of the standard 90° V-notch weir is determined directly by the head on the bottom of the V-notch.

As the stream passes over the weir, the top surface curves downward. This curved surface, or drawdown, extends upstream a short distance from the weir notch. The head must be measured at a point on the water surface in the weir pond beyond the effect of the drawdown. This distance should be at least four times the maximum head on the weir, and the same gauge point should be used for lesser discharges. A staff gauge having a graduated scale with the zero placed at the same elevation as the weir crest is usually provided for the head measurements. Figure 3-2 illustrates a standard staff gauge adopted by the Bureau.

After the head is determined, the rate of flow, or discharge, may be found by referring to the tables described in succeeding paragraphs. These tables are for free-flow conditions and are applicable only to weirs installed in accordance with the requirements for standard contracted weirs set forth in the preceding paragraphs. Many of the values shown in these tables were determined experimentally, the remainder were computed from the accepted formulas. Table 3-1 gives the maximum and minimum capacities of some standard weirs.

38

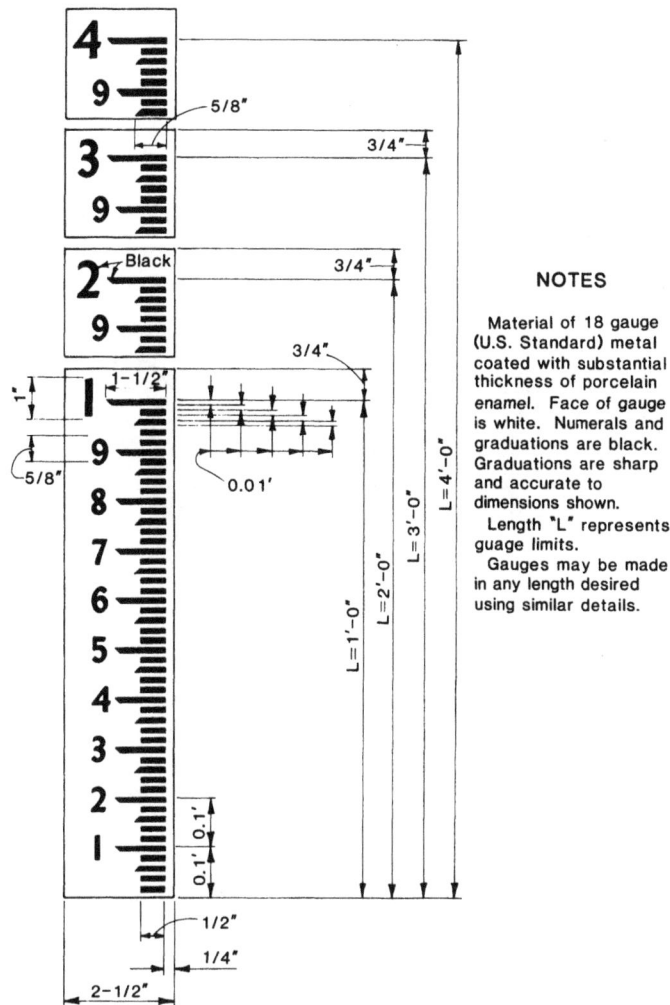

Figure 3-2.—Standard staff gauge. 103-D-855.

NOTES

Material of 18 gauge (U.S. Standard) metal coated with substantial thickness of porcelain enamel. Face of gauge is white. Numerals and graduations are black. Graduations are sharp and accurate to dimensions shown.

Length "L" represents guage limits.

Gauges may be made in any length desired using similar details.

e. *Discharge Equations.*—There are several well-known equations used to compute the discharge over 90° V-notch weirs. The most commonly used are the Cone and Thomson equations. The Cone equation is considered by many authorities to be the most reliable for conditions generally encountered with small weirs:

$$Q = 2.49 \ H^{2.48} \ \text{(inch-pound units)} \qquad (1)$$
$$Q = 1.34 \ H^{2.48} \ \text{(SI-metric)}$$

where:

Q = discharge over weir in cubic feet per second (cubic meters per second), and
H = head on weir in feet (meters).

Ordinarily, V-notch weirs are not appreciably affected by velocity of approach. If the weir is installed with complete contraction, the velocity of approach will be low.

Table 3-2 shows discharges for the standard 90° contracted V-notch weir, without velocity of approach, computed from the Cone formula for a range of heads ordinarily used in measuring small flows.

f. *Maintenance.*—For best operating conditions, the weir structure should be set in a straight reach of the channel and perpendicular to the line of flow. The weir crest must be level and the bulkhead plumb.

The level of the crest should be checked periodically, and should also be checked with reference to the elevation of the zero of the staff gauge. Inspection should also be made to determine whether there is leakage around the weir and, if such leakage exists, the structure should be repaired.

Care must be taken to avoid damaging the weir notch itself. Small nicks and dents can reduce the accuracy of an otherwise good weir installation. Any nicks and dents that do occur should be carefully dressed with

39

Table 3-1. – Maximum and minimum capacities of standard weirs[1].

Length, feet	Contracted Rectangular		Suppressed Rectangular		Cipolletti	
	Maximum	Minimum	Maximum	Minimum	Maximum	Minimum
1.0	0.590	0.286	0.631	0.298	0.638	0.301
1.5	1.65	.435	1.77	.447	1.79	.452
2.0	3.34	.584	3.65	.596	3.69	.602
2.5	5.87	.732	6.30	.744	6.37	.753
3.0	9.32	.881	10.0	.893	10.1	.903
3.5	13.8	1.03	14.8	1.04	15.0	1.05
4.0	19.1	1.18	20.4	1.19	20.6	1.20
4.5	25.7	1.33	27.5	1.34	27.8	1.35
5.0	33.5	1.48	36.0	1.49	36.4	1.51
5.5	42.3	1.63	45.3	1.64	45.8	1.66
6.0	52.7	1.78	56.6	1.79	57.2	1.81
7.0	77.4	2.07	82.9	2.08	83.8	2.11
8.0	108.5	2.37	116.2	2.38	117.5	2.41
9.0	145.3	2.67	155.9	2.68	157.6	2.71
10.0	188.8	2.97	202.4	2.98	204.6	3.01
12.0	298.4	3.56	320.0	3.57	323.6	3.61
14.0	439.1	4.16	470.4	4.17	475.6	4.21
16.0	612.0	4.75	656.5	4.76	663.8	4.82
18.0	822.4	5.35	882.0	5.36	891.8	5.42

(a) Capacities shown in cubic feet per second.

Length, meters	Contracted Rectangular		Suppressed Rectangular		Cipolletti	
	Maximum	Minimum	Maximum	Minimum	Maximum	Minimum
0.30	0.016	0.008	0.017	0.008	0.018	0.008
.40	.033	.010	.036	.011	.036	.011
.50	.058	.013	.063	.014	.063	.014
.75	.161	.020	.172	.020	.174	.020
1.00	.330	.027	.354	.027	.358	.027
1.25	.577	.033	.618	.034	.625	.034
1.50	.910	.040	.975	.041	.986	.041
1.75	1.338	.047	1.433	.047	1.449	.048
2.00	1.868	.054	2.001	.054	2.024	.055
2.50	3.263	.067	3.496	.068	3.535	.068
3.00	5.148	.081	5.515	.081	5.577	.082
3.50	7.568	.094	8.108	.095	8.199	.096
4.00	10.567	.108	11.322	.108	11.448	.109
4.50	14.185	.121	15.199	.122	15.367	.123
5.00	18.460	.135	19.779	.135	19.998	.137
5.50	23.427	.148	25.100	.149	25.379	.150
6.00	29.120	.162	31.200	.162	31.546	.164

(b) Capacities shown in cubic meters per second.

[1]Limits follow the prescribed practice of $h > 0.2$ foot (0.06 m) and $h < L/3$.

Table 3-2 – Discharges of 90° V-notch weirs.

(a) Inch-pound units, computed from the equation $Q = 2.49\,H^{2.48}$

Head, feet	Discharge, ft³/s	Head, feet	Discharge, ft³/s	Head, feet	Discharge, ft³/s
0.20	0.046	0.55	0.564	0.90	1.92
.21	.052	.56	.590	.91	1.97
.22	.058	.57	.617	.92	2.02
.23	.065	.58	.644	.93	2.08
.24	.072	.59	.672	.94	2.13
.25	.080	.60	.700	.95	2.19
.26	.088	.61	.730	.96	2.25
.27	.096	.62	.760	.97	2.31
.28	.106	.63	.790	.98	2.37
.29	.115	.64	.822	.99	2.43
.30	.125	.65	.854	1.00	2.49
.31	.136	.66	.887	1.01	2.55
.32	.147	.67	.921	1.02	2.61
.33	.159	.68	.955	1.03	2.68
.34	.171	.69	.991	1.04	2.74
.35	.184	.70	1.03	1.05	2.81
.36	.197	.71	1.06	1.06	2.87
.37	.211	.72	1.10	1.07	2.94
.38	.226	.73	1.14	1.08	3.01
.39	.240	.74	1.18	1.09	3.08
.40	.256	.75	1.22	1.10	3.15
.41	.272	.76	1.26	1.11	3.22
.42	.289	.77	1.30	1.12	3.30
.43	.306	.78	1.34	1.13	3.37
.44	.324	.79	1.39	1.14	3.44
.45	.343	.80	1.43	1.15	3.52
.46	.362	.81	1.48	1.16	3.59
.47	.382	.82	1.52	1.17	3.67
.48	.403	.83	1.57	1.18	3.75
.49	.424	.84	1.61	1.19	3.83
.50	.445	.85	1.66	1.20	3.91
.51	.468	.86	1.71	1.21	3.99
.52	.491	.87	1.76	1.22	4.07
.53	.515	.88	1.81	1.23	4.16
.54	.539	.89	1.86	1.24	4.24

(b) SI-metric units, computed from the equation $Q = 1.34\,H^{2.48}$

Head, meters	Discharge, m³/s	Head, meters	Discharge, m³/s	Head, meters	Discharge, m³/s
0.055	0.0010	0.205	0.0264	0.355	0.1029
.060	.0013	.210	.0280	.360	.1065
.065	.0015	.215	.0297	.365	.1102
.070	.0018	.220	.0314	.370	.1140
.075	.0022	.225	.0332	.375	.1179
.080	.0026	.230	.0351	.380	.1218
.085	.0030	.235	.0370	.385	.1258
.090	.0034	.240	.0390	.390	.1299
.095	.0039	.245	.0410	.395	.1341
.100	.0044	.250	.0431	.400	.1384
.105	.0050	.255	.0453	.405	.1427
.110	.0056	.260	.0475	.410	.1471
.115	.0063	.265	.0498	.415	.1516
.120	.0070	.270	.0522	.420	.1562
.125	.0077	.275	.0546	.425	.1608
.130	.0085	.280	.0571	.430	.1655
.135	.0094	.285	.0597	.435	.1703
.140	.0102	.290	.0623	.440	.1752
.145	.0112	.295	.0650	.445	.1802
.150	.0122	.300	.0678	.450	.1853
.155	.0132	.305	.0706	.455	.1904
.160	.0143	.310	.0735	.460	.1957
.165	.0154	.315	.0765	.465	.2010
.170	.0166	.320	.0796	.470	.2064
.175	.0178	.325	.0827	.475	.2119
.180	.0191	.330	.0859	.480	.2175
.185	.0204	.335	.0891	.485	.2231
.190	.0218	.340	.0925	.490	.2289
.195	.0233	.345	.0959	.495	.2347
.200	.0248	.350	.0994	.500	.2406

a fine-cut file or stone, stroking only in the plane of the weir upstream face, plane of the weir crest or sides, or plane of the chamfers. Under no circumstances should the upstream corners of the notch be rounded or chamfered, nor should any attempt be made to remove completely an imperfection that would change the shape of the weir opening. Instead, only those portions of the metal that protrude above the normal surfaces should be removed.

   g. *Data Processing and Review.*—Field data consisting of head measurements are entered on field data forms as shown on figure 3-3. The data are then transmitted to the Denver Office in units of gallons per minute (liters per minute) or cubic feet per second (cubic meters per second). Upon arrival, the data are

---

FRYINGPAN–ARKANSAS PROJECT
PUEBLO DAM

Inspection Gallery Seepage Water Measurements

(90° V-Notch Weirs)

Date ___1-13-84___   Observer ___MUSGROVE___

Weir No.	Head (feet)	Discharge (gal / min)
1	0.38	101
2	0.32	66
3	0.25	36
4	0.23	29
5	0.20	
6	0.34	
7	0.51	210

Total (gal/min)___442___

Reservoir Elevation ___4877.80 ft___

Remarks:
   WEIR NO. 7 IS LOCATED IN THE SUMP IN BUTTRESS 16, AND THE FLOW MEASURED IN THIS WEIR INCLUDES THE FLOW MEASURED BY WEIRS 5 AND 6.

Figure 3-3.—Field data form for weir readings.

42

processed and reviewed as described in section 1.25. Many of the calculations are now performed automatically by a computer program. Data plots are then prepared as shown on figure 3-4.

3.3 Velocity Meters.—a. *Usage.*—Many types of velocity meters are available commercially, and their methods of operation vary from the Pitot tube principal to propeller-type devices, accoustic flowmeters, and electromagnetic current indicators. Most of these devices can be used for measuring flow in pipes or open channels. A relatively new device used by the Bureau is the portable velocity meter, which can be used for measuring water velocity.

b. *Description of Devices.*—The portable velocity meter probe operates on elementary electromagnetic principles in that a conductor moving through a magnetic field will have an induced voltage. For the velocity meter shown on figure 3-5, a signal is generated and sent to an electromagnet within the probe, which creates a magnetic field. The conductor is the water into which the probe is immersed. As water flows through the magnetic field, a voltage is generated in the water in the vicinity of the electrodes which sense the voltage. The sensed voltage is then transmitted through the cable to the surface unit which amplifies and conditions the signal, and displays the results as a velocity measurement. The polarity and magnitude of this signal is directly proportional to the direction and velocity of the water. By knowing the channel dimensions and depth of flow, the quantity of flow may be determined.

The device consists of a probe, extension rods, and a surface readout unit, which are all connected by a cable. The unit is battery powered, lightweight, and easily portable. Calibration tests by Bureau laboratories indicate an accuracy within 5 percent at low flows of about 0.1 ft³/s (0.003 m³/s) and within 1 percent at higher flows of about 2.0 ft³/s (0.06 m³/s).

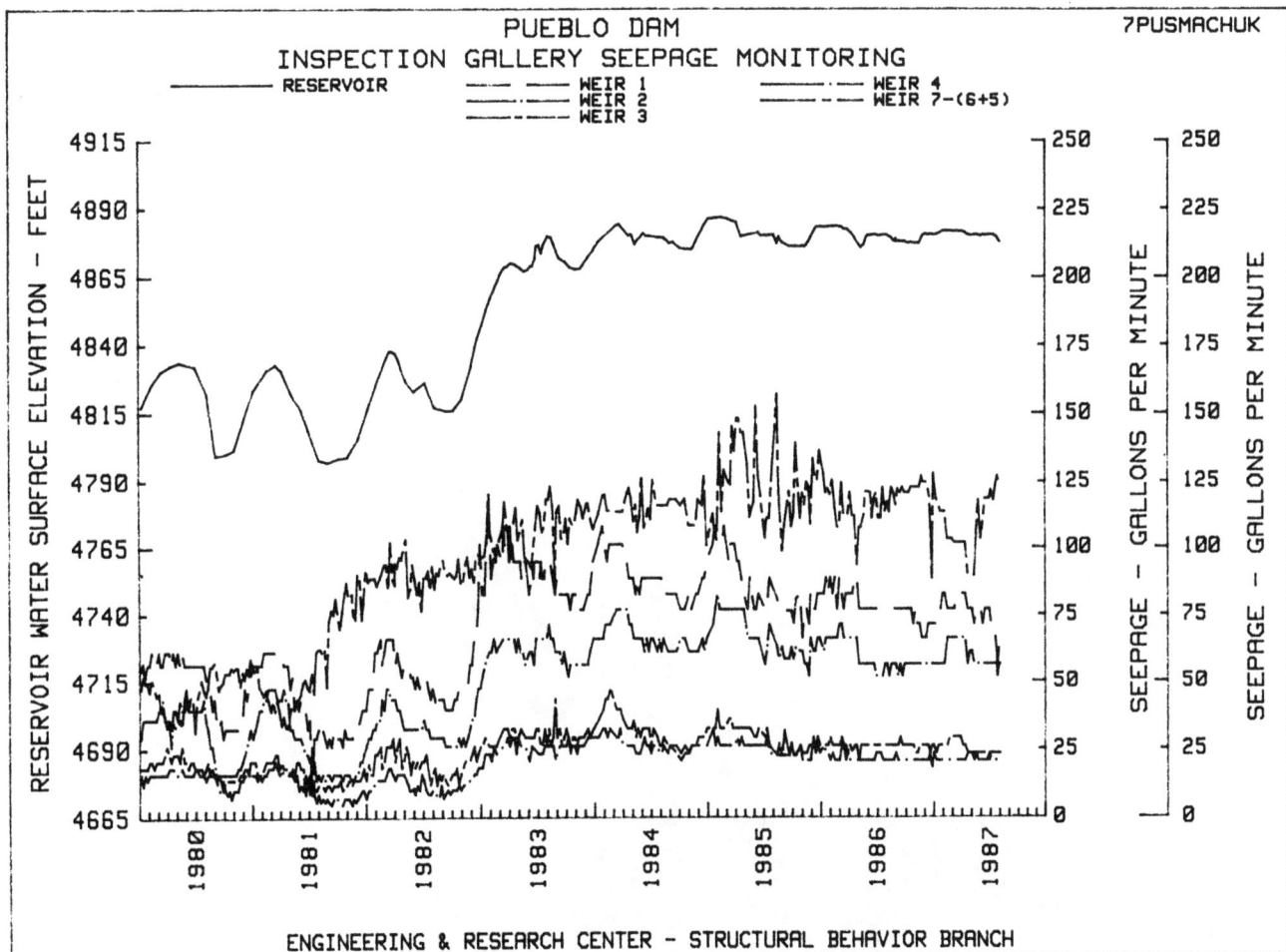

Figure 3-4.—Data plots on weir discharges.

43

c. _Installation of Devices._—No installation is required for the portable velocity meter, the device is merely lowered into the flowing water and readings made.

d. _Monitoring Procedures._—The velocity meter is lowered into the flowing water with the device switches set at "Normal" and "On". The water velocity is read directly from the meter. For example, assume a 12-inch (305-mm) diameter pipe flowing 6 inches (152 mm) deep at a velocity of 2 ft/s (0.6 m/s). Using a table similar to table 3-3, the depth of flow, to the nearest ⅛ inch (3 mm), in the pipe is found along the left side of the table and velocity along the top. For this example, a value of 0.509 million gallons per day (1.93 million liters per day) is determined for the flow. Values of velocity not equal to whole numbers are entered as whole numbers and as fractions of whole numbers, and then added together. For example, using the previous assumptions and a measured velocity of 2.5 ft/s (0.8 m/s), the 0.509 million gallons per day is added to 0.127

Figure 3-5.—Portable velocity meter. P801-D-81190.

44

million gallons per day, which is 0.1 the flow in the 5-ft³/s (0.14-m³/s) column, for a total flow of 0.636 million gallons per day (2.41 million liters per day).

Open channel or pipe flow may be determined by measuring the velocity and computing the cross-sectional area of the channel or pipe to the depth of flow. Then, the flow may be computed by multiplying the measured water velocity times the area.

3.4 Calibrated Container Devices.—The most basic method of monitoring quantity of flow from drains consists of merely catching a known quantity of water in a calibrated container and measuring the amount of time required to do so. However, such a method is normally reserved for relatively low flow conditions.

This method requires that the drain water be either flowing through a pipe with an exposed end or that the channel the water is flowing in can be constructed with a vertical drop and an overhang. Such requirements are necessary to be able to place the container in a position to catch the water.

Calibrated containers may be of any size, but commonly used sizes are 0.5, 1.0, or 2.0 gallon (1.9, 3.8, or 7.6 L) in capacity for convenience of handling. When catching the water, the container is held in a position so as to catch the total flow, and the time in minutes, or fraction thereof, required to fill the container is noted. For example, if a 1.0-gallon (3.8-L) container is filled in exactly 1 minute, the flow rate is 1 gal/min (3.8 L/min). Similarly, a 0.5-gallon (1.9-L) container filled in exactly 40 seconds would indicate a flow rate of 0.5/(40/60), or 0.75 gal/min (2.84 L/min).

The data are recorded on a field data form as shown on figure 3-6 and transmitted to the Denver Office for processing as described in section 1.25.

SEEPAGE MONITORING
(CONTAINERS)

Modified_____ Observer(s) _____ Sheet_____of_____
for_____Dam   Reservoir Elev._____ Date_____
                    Tailwater Elev._____

MEASURING DEVICE NO.	CONTAINER SIZE	TIME	READING** (GPM)	PRECIPITATION	REMARKS

Report precipitation for a week prior reading.

**Report even if reading is dry. Dry reading is
assumed as a zero reading.*

Figure 3-6.—Field data form for a calibrated container.

Chapter 4

INTERNAL MOVEMENT (EMBEDDED) MEASUREMENT DEVICES

A. Usage and Types of Instruments

4.1 General.—Many internal movement measuring devices have been developed over the years. This chapter will discuss only those devices that are installed within the dam, foundation, or abutments. Surface devices and external measurement monuments will be discussed in chapter 5.

It is desirable to determine the amount and rate of several types of movements in and adjacent to a concrete dam. These movements may be classed as vertical, horizontal (translational), or rotational. Vertical movements are commonly the result of consolidation of foundation soils or rock due to the mass of the dam and reservoir. Horizontal or translational movements refer to movements of a dam that are approximately perpendicular or parallel to the axis of the dam. These movements may take the form of simple translational movement of the mass of a dam in the downstream direction due to high water pressures of the reservoir or temperature fluctuations, or they may be due to lateral spreading of the foundation soil or rock. Rotational movements may occur because of low shearing strength of the foundation soil or rock. In this case, the lower portion of the dam moves outward with respect to the center of the dam. In general, gravity dams may move up and down, upstream or downstream, and rotate in the upstream-downstream direction, while arch dams may translate or rotate in all three axes.

Instruments for measuring movements in a dam are installed for the life of a structure. Therefore, to give satisfactory performance over a long period, the devices should be relatively resistant to corrosion, readily adaptable to site conditions, durable, relatively easy to install, readily procurable, of simple operating design and, of course, be capable of yielding the information desired with the precision and accuracy required. Recording instruments should be simple and direct reading, either mechanical or electrical, and able to be operated correctly by trained personnel. Over a period of several years, results from these measurements will indicate the range of movement of a structure during the various loading conditions of temperature and water pressures.

4.2 Types of Instruments.—Devices used to measure internal movements include plumblines, extensometers, tiltmeters or inclinometers, strain meters, and joint meters. Each of these types will be discussed in detail in succeeding sections of this chapter, along with their advantages and limitations as perceived by Bureau experience.

B. Currently Used Devices

4.3 Plumblines.—a. *Usage.*—Plumblines represent a convenient and relatively simple method of measuring the manner in which a dam moves due to applied reservoir water pressures and temperature changes. In early Bureau dams where elevator shafts were provided, plumblines were located in the elevator shafts. That practice generally proved to be unsatisfactory because of mechanical- or airflow-induced vibrations in the plumbline wire.

For the more recently built Bureau dams, the plumblines have been suspended in vertically formed wells that extend from the top of the dam to near the foundation, generally at three or more locations in the dam. Wherever feasible, reading stations are located at intermediate elevations, as well as at the lowest possible elevation, to measure the deflected position of the dam section over the full height of the structure. A typical installation is shown on figure 4-1. Plumblines may also extend into the foundation in drill holes.

b. *Description of Devices.*—Plumbline installations of two general types have been used, the weighted plumbline and the float-suspended or inverted plumbline. The installations consist of a formed shaft, suspension assembly, wire, plumb bob, dashpot, reading stations and, for float-suspended types, a tank and float assembly.

(1) *Shaft.*—The shafts or wells are usually 1 foot (0.3 m) in diameter, and are constructed to within 0.5 inch (13 mm) of plumb as the dam increases in elevation. In some dams, pipe or casing has been used

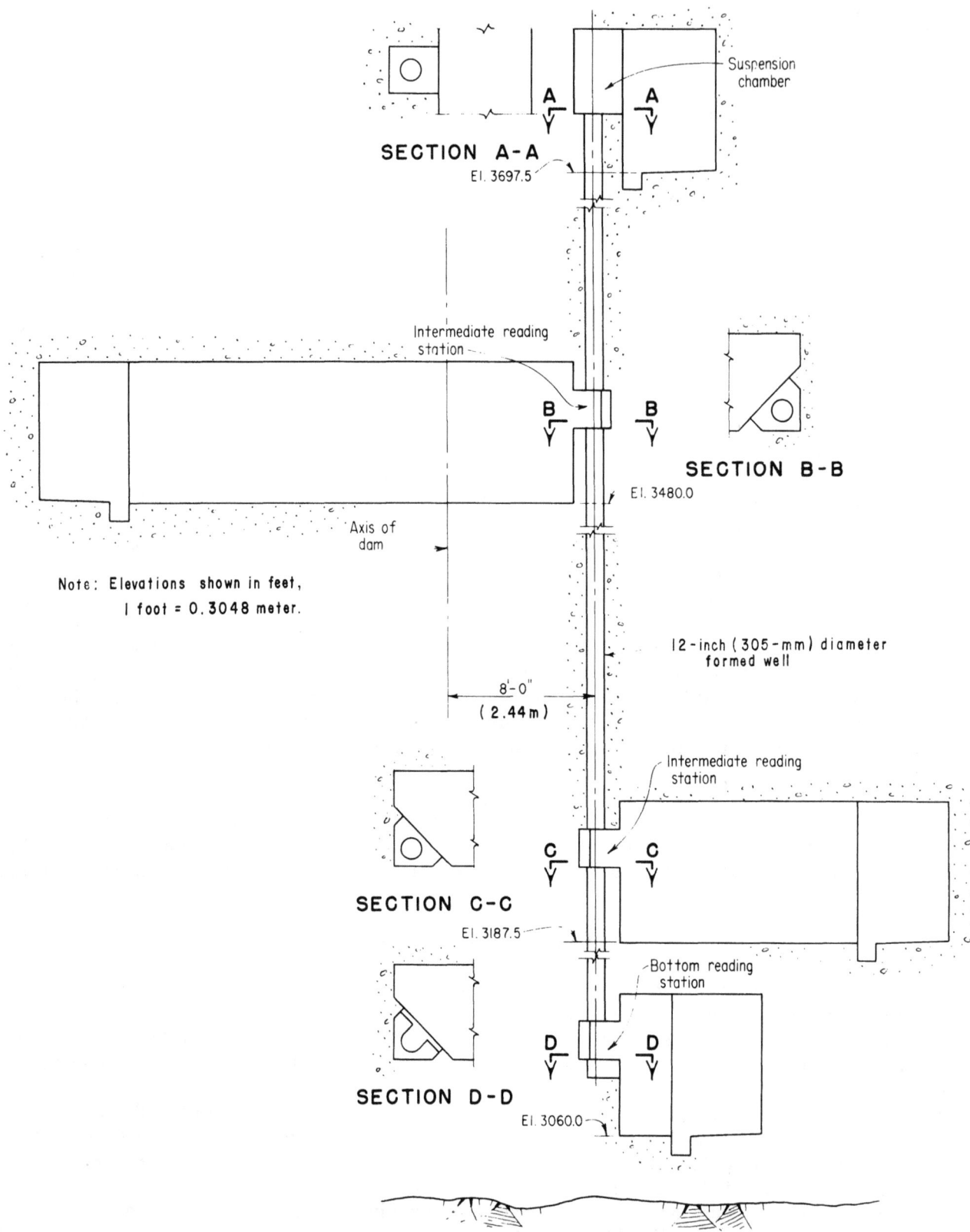

SECTION A-A

EI. 3697.5

Suspension chamber

Intermediate reading station

SECTION B-B

EI. 3480.0

Axis of dam

Note: Elevations shown in feet, 1 foot = 0.3048 meter.

12-inch (305-mm) diameter formed well

8'-0"
(2.44 m)

Intermediate reading station

SECTION C-C

EI. 3187.5

Bottom reading station

SECTION D-D

EI. 3060.0

Figure 4-1.—Typical plumbline installation. 288-D-3090.

to form the shaft and has been left in place, while in other dams, the shafts have been formed using slip-forms. Pipe casing left in place has, in some cases, resulted in problems with corrosion and rust flaking.

(2) *Suspension Assembly.*—For the conventional plumbline, the wire is suspended at the upper end of the pipe shaft using a collet and nut inserted in the center of a steel suspension assembly placed over the pipe opening. Alternatively, other devices, such as a pulley and locking spool, could be used to suspend the plumb bob. Watertight gaskets and sealing materials are used to prevent entry of water into the shaft and to reduce corrosion of the plumbline apparatus.

(3) *Wire, Plumb Bob, and Dashpot.*—The plumbline wire is usually of stainless steel or other high-strength, corrosion-resistant steel wire that is $1/_{64}$ inch (0.4 mm) in diameter. A 15- to 25-pound (6.1- to 11.3-kg) plumb bob is normally used, and is also constructed of a corrosion-resistant metal. To damp the pendulum action of the wire and plumb bob and to minimize local vibrations of the wire, an oil-filled container is provided at the lowest reading station for immersion of the plumb bob. A noncorrosive 20-W motor oil should be used. This dashpot must be at least 4 inches (102 mm) larger in radius than the plumb bob to allow for free deflection. When moisture or condensation is excessive, a deflecting metal cover may be clamped around the wire a short distance above the dashpot to prevent water contamination of the oil.

(4) *Reading Stations.*—The reading stations on a plumbline shaft are located in the galleries at different elevations in the dam. A doorframe is set in the concrete of the gallery wall at each reading station, and doors seating against rubber seals are provided as closures, see figure 4-2. The doors of the reading station should remain closed and locked to prevent disturbance to the plumbline.

If possible, reading stations are oriented so that measurements may be made in the directions of anticipated movements, thereby avoiding the need for trigonometric resolution. In some older dams, orientation of the reading stations requires that measurements be made at 45° to the direction of dam movement, thus necessitating computations to determine radial and tangential components of movements.

Measurements of deformation are made with a micrometer slide device having either a peep sight or microscope for viewing. The measured movements indicate deformation of the structure with respect to the plumbline suspension point. The readings are also readily adaptable to automated reading systems.

(5) *Tank and Float Assembly for Inverted Plumbline.*—A few recently installed plumbline installations are float suspended, using antifreeze in a tank near the top of the dam, with a float holding the wire. In this type of installation, the lower end of the plumbline is fixed near the bottom of the dam and movements are observed near the top and at intermediate stations. Figure 4-3 shows a installation of this type.

c. **Advantages and Limitations.**—Advantages of plumbline devices include high accuracy in movement measurements and, when observed over several years, will furnish information regarding the general elastic behavior of the entire structure and foundation. These data also provide a means for determing the elastic shape of the deflected structure and, with precise alignment data, provide for the estimation of the amount of translation of the structure.

Limitations of conventional plumbline usage include the requirement of trained observers for securing the readings, possible corrosion of metal components in the system, and the fact that no readings can be obtained until completion of the structure to its full height. With inverted plumblines, measurements can be made during construction.

d. **Installation Procedures.**—(1) *Shaft and Reading Station.*—The process of forming the reading station recess is a routine construction procedure and is included with other form work in the construction contract. The plumbline shaft is made by installing lengths of metal pipe vertically from below the lowest station up to the suspension point. Spiral welded pipe or drill hole casing pipe have been found satisfactory for this purpose although problems with rust flaking have been experienced. Special care in alignment and bracing during concrete placement is necessary to ensure plumbness of the pipe sections and to ensure a clear projected net opening approaching the full size of the pipe.

(2) *Suspending the Plumb Bob.*—It is important that the permanent plumbline is located at or near the center of the net opening of the shaft. A method that has been used successfully is as follows: center point may be determined by suspending a surveying plumb bob on a string or cord of a length sufficient to reach from the top of the shaft to the lowest reading station. A sheet of cardboard is then secured in place directly beneath lower end of shaft, and a pattern of the clear projected net opening of the pipe is established by marking the positions of the plumb bob as the suspension cord is moved around the periphery of the shaft at the top of the shaft. The plumb bob should be allowed to come to rest at each of the 8 to 10 points required to define the clear opening pattern. With the cardboard still in place, the center of the pattern is established

Figure 4-2.—Typical plumbline reading station. P801-D-81191.

within ¼ inch (6 mm) and is projected to the top of the shaft using the temporary plumbline. The permanent plumbline is then set as close to the center as practicable.

Installing the permanent plumbline is not a difficult operation, but must be accomplished with care to avoid kinks or twists in the wire. A method that can be used is as follows: the wire, which has previously been wound on a spool in a reeling rack, is threaded through the collet in the plug, through suspension assembly, and then attached to the plumb bob. On some plumb bobs, the cap may be removed and the connection made by threading the wire through the hole in the center of the cap which contains a cone-shaped recess. The wire is twisted around a short nail, and then hot solder is poured in to cover the nail and fill the cup. After solder has cooled, the cap is lowered through the shaft to the lower reading station. When the cap is at the level of the damping pot, cap is screwed onto plumb bob, lowered into damping pot, and sufficient oil added to cover plumb bob. The freely suspended bob is then adjusted to an elevation just below the oil surface and the wire permanently fixed at the suspension point.

For the tank and float-suspended plumbline, the center of the shaft is found as before, and the wire is fixed at the bottom and fastened to the center of the float device at the top. The tank at the top can be moved to allow for centering the float device.

(3) *Reading Stations.*—The position of the four bars (two microscope mounts and two reference bars) attached to the bar support plates in the reading stations is dependent on the following:

- Maximum expected movement of plumbline
- Mechanical working distance of microscope (distance from front of objective lens mount to object plane).

50

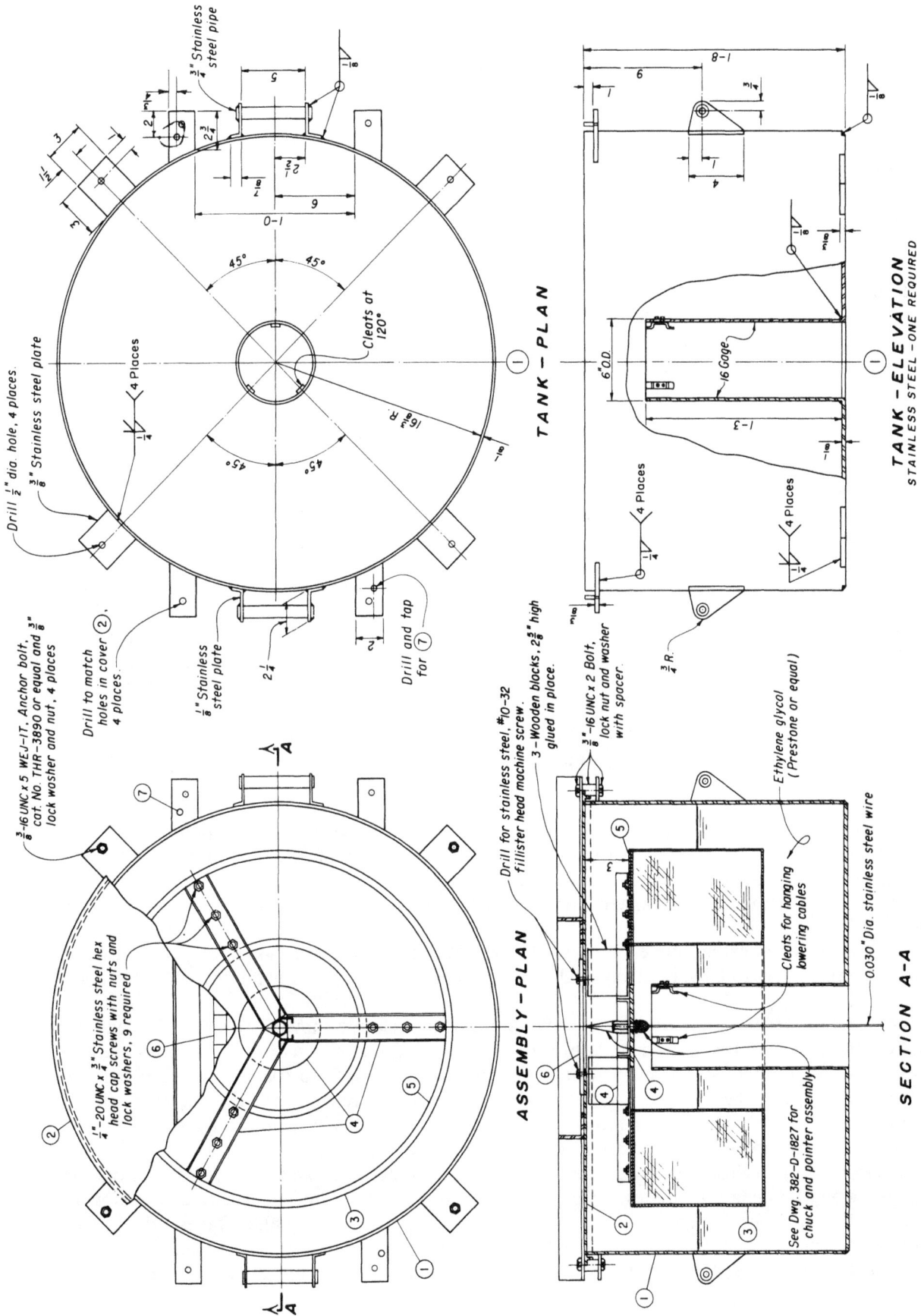

Figure 4-3.—Details of a float-suspended plumbline. From 382-D-1826.

51

- Length of telescope drawtube travel (external range of rack and pinion movement of objective lens mount).
- Offset dimensions of micrometer support clamp.

The positions are established by consideration of the extreme positions of the plumbline during maximum deflection due to reservoir load and the estimated annual temperature deflection cycle. Both the annual temperature deflection cycle and the maximum load deflection cycle are superimposed upon the estimated permanent thermal deflection to obtain the maximum possible range normal to the axis of the dam. Transverse movements are generally small and difficult to predict, so an allowance of about 10 percent of the total estimated movement normal to the axis of the dam is provided in the lateral direction. The reference bars and slides are then mounted to allow such movement to be accurately measured.

    e. Maintenance.—A light film of oil should be maintained on the reference and support bars to provide protection against corrosion. It is necessary to wipe away this oil immediately prior to taking readings. After readings are obtained, replace the light film of oil. The oil in the damping pot should be maintained at the proper level.

When not in use, micrometer slides, micrometers, and microscopes should be kept in their carrying cases and stored in a safe, dry place. This apparatus must be handled with care at all times. One or two drops of light lubricating oil may be applied to the exposed screw threads and sliding surfaces of the micrometer at intervals of several months. It is suggested that a soft brush and tissue paper be used for removing dust, as necessary.

The inverted system should be inspected for alignment and for proper fluid level in tank.

    f. Data Collection and Processing.—The micrometer, micrometer slide, and microscope are precision instruments and should be used in a manner consistent with good laboratory practice. All microscope readings should preferably be made by the same person; however if that is not practicable, a new observer must be thoroughly trained in instrument usage.

To take a reading, the micrometer, slide assembly, and microscope are fixed into a position such that the reading will fall within the range of the microscope. All clamps are then fixed, and the microscope always started from the left-hand side. The screws are turned so that the objective lens approaches the plumbline wire from the left side. Continue turning the screws until the wire is precisely sighted, and then a reading is taken. If the screws are turned to a point where the lens is focused to the right side of the line, the microscope must be returned some distance to the left and the line approached again. The screws should never be reversed just prior to making a reading.

The readings are recorded on a field data form as shown on figure 4-4. The data forms are forwarded to the Bureau's Denver Office for processing and analysis as discussed in section 1.25. Plots of movements are prepared as shown on figure 4-5.

    4.4 Extensometers.—Extensometers may be divided into two general types, the single-point and the multiple-point. Both of these types are used to measure movement of one portion of a dam relative to another portion, movement of the dam relative to the foundation, or movement of one portion of the foundation relative to another portion. Additional extensometer types are used to measure movements on dam surfaces and will be discussed in chapter 5.

    a. Usage.—Internal extensometers are usually installed in uncased drill holes, and are suitable for installation vertically, horizontally, or at any angle. Extensometers used by the Bureau include the rod and wire types.

    b. Advantages and Limitations.—Extensometers measure relative movement along the length of the installation; therefore, relative movement between any portion of a dam or its foundation may be measured. However, most extensometers have an operating range of 2 to 4 inches (51 to 102 mm), so movements should not exceed this range. If the movement should exceed the operating range of the extensometer, the reading head must be reset and a constant added to all future readings. With proper maintenance and care in reading, extensometers provide excellent information on relative movement.

    c. Description of Devices.—Extensometers are designed to measure axial displacement of one or more fixed points along the length of the extensometer. The rod-type extensometer usually consists of multiple anchors installed at different depths in a drill hole. Aluminum rods inside hollow tubes extend from each anchor to a reference head at the collar of the hole. All measurements of movements are made at the reference head. As an anchor moves, the rod attached to that anchor also moves and the amount of movement is

UNITED STATES
DEPARTMENT OF THE INTERIOR
BUREAU OF RECLAMATION
COLUMBIA BASIN PROJECT - WASHINGTON

GRAND COULEE THIRD POWER PLANT

FOREBAY DAM

DAM DEFLECTOMETER MEASUREMENTS

(PEEP SIGHT VIEWER)

	I.D.	MO.	DAY	YR.
INITIAL DATE (1–10)	F.B	0.7	0.7	7.2
READING DATE (11–20)	F.B	0.5	2.7	8.7

	ELEVATION (21–)	BLOCK NO. (–30)
Sign +	1.1.2.7.5.0	1.0.2

CHECK ONE WHEN READING INSTRUMENT:

☒ READER FACING RESERVOIR

☐ READER FACING AWAY FROM RESERVOIR

INSTRUMENT READINGS	
SEQUENCE	WIRE
1ST. READING	(1) [31–35] 8.0.6.7.0
2ND. READING	(2) [36–40] 2.5.0.8.0
3RD. READING	(3) [41–45] 2.5.0.8.0
4TH. READING	(4) [46–50] 8.0.6.7.0
5TH. READING	(5) [51–55] 8.0.6.7.0

RES. ELEV. 1277.69 RECORDER A.W., J.C. CHECKED BY J.C. TIME 0800 TO 1100

COMMENTS:

(SEE INSTRUCTIONS ON SEPARATE SHEET)

Figure 4-4.—Field data form for plumbline readings. 1222–D–2964.

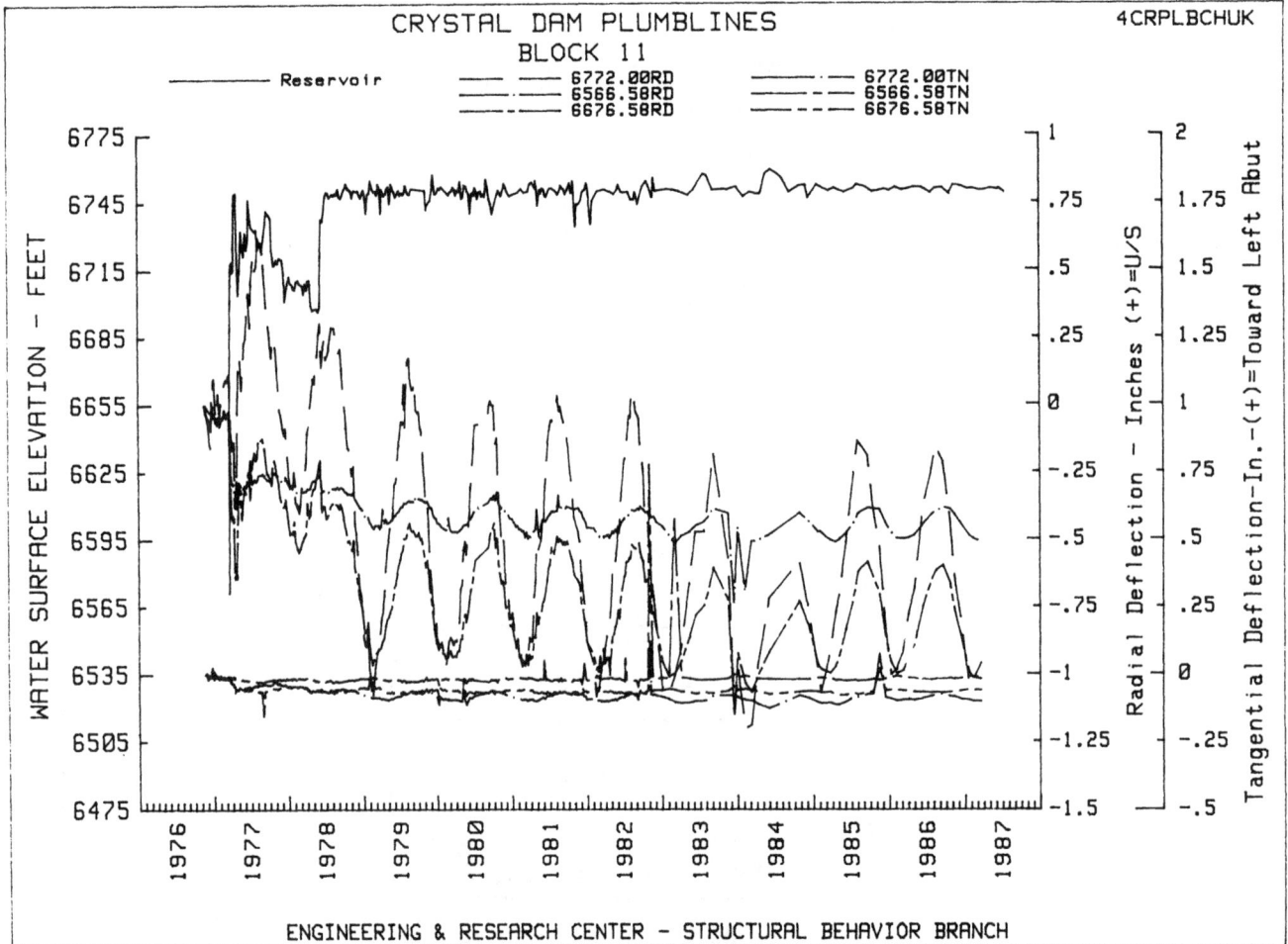

Figure 4-5.—Data plots of a plumbline installation.

measured at the reference head. Theoretically, depending on the size of the drill hole, as many rods and anchors may be installed as desired, but normally only 5 to 10 anchors are installed per hole.

Movement of the anchors may be measured either mechanically with depth gauges or electrically with linear potentiometers. Various types of anchors are available and most operate by radial expansion to lock the anchor in place. Figure 4-6 shows a rod-type extensometer; this device can measure up to 4 inches (102 mm) of movement before resetting of the potentiometer is required.

Wire-type extensometers (fig. 4-7) use one or more tension wires instead of rods extending from an anchor or anchors to the reference head. Multipoint extensometers normally have six or eight anchors connected to each reference head. An older type of extensometer had the sensor head end of the wire connections containing a stainless steel cantilever device for each wire. Changes in distance between the anchors and the head were sensed by deflections in the cantilevers. Movements of the cantilevers were detected through the use of strain gauges, and were checked or calibrated using a depth micrometer or dial gauge. A high degree of accuracy is available with this device; however, only ± 0.30 inch (± 7.6 mm) of movement is allowed before the wires or head must be reset.

d. Installation Procedures.—Upon completion and cleaning of the borehole, the anchor string with head packer attached is inserted into the hole and the anchors fixed in place if they are the expandable type. A string that is to be grouted and is not equipped with fixable anchors must have only the bottom portion grouted, or otherwise securely anchored, to prevent "floating" when grout is pumped into the hole. When bottom anchor is fixed, the rest of the hole is grouted full using a thin sand-cement grout. The measuring head is then installed, ensuring that appropriate rods or wires are attached to the correct transducer or cantilever.

54

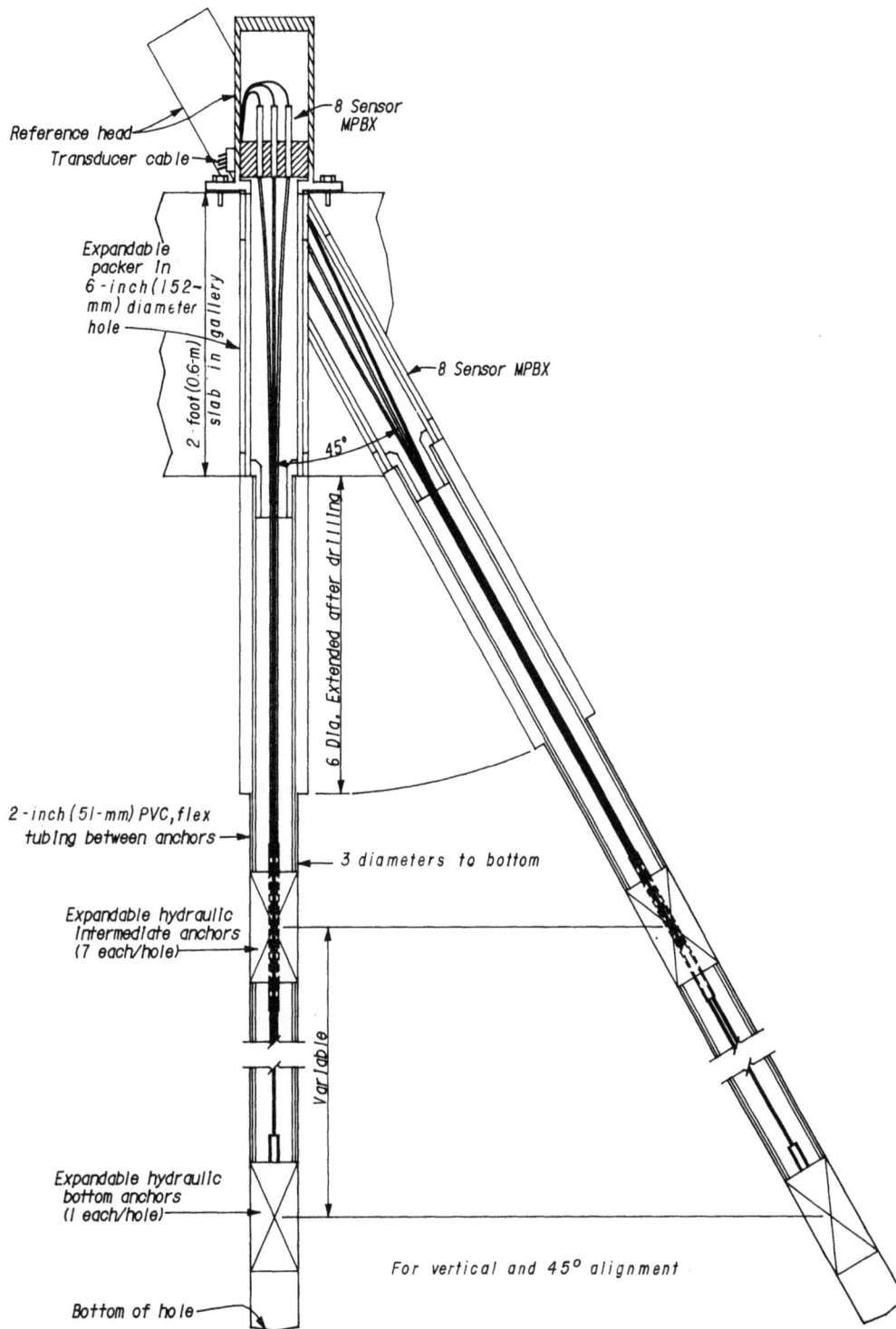

Figure 4-6.—Details of a rod-type extensometer. 963-D-128.

e. Monitoring Procedures.—Extensometers are read on a scheduled periodic basis. Either mechanical readings using depth gauges or electrical readings using portable readout devices are made. Most electrical readout units convert the voltage readings from the electrical sensor to a direct display reading, which is usually in inches. Corrections for spring tension, wire stretch, and temperature are included to provide a final value. Differences between initial and current readings provide the actual movement data.

f. Maintenance.—The reference heads must be kept free of dust, grit, and moisture to ensure proper, long-term operation. Occasionally, if significant movement has occurred, it is necessary to reset one or more

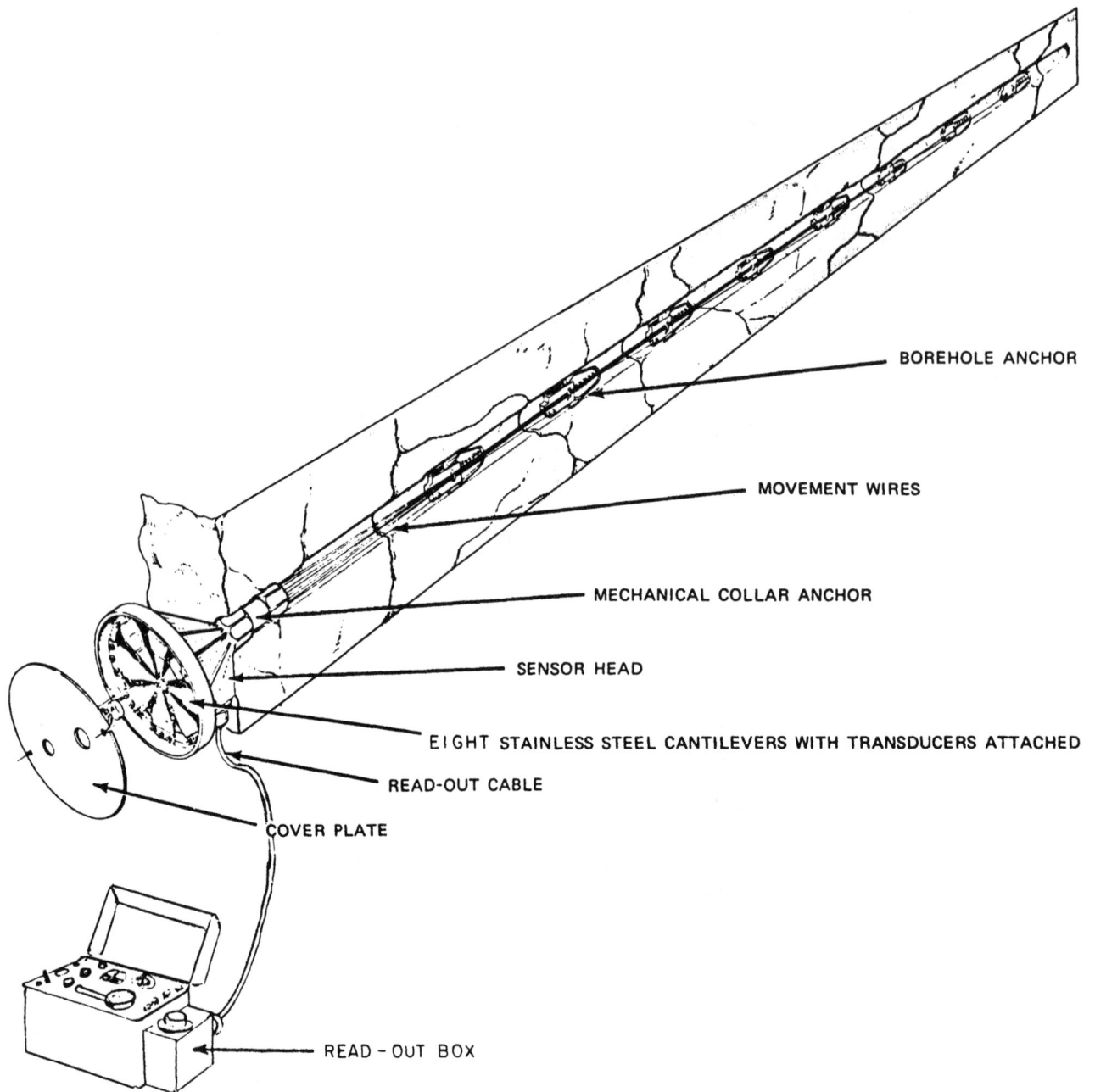

Figure 4-7.—Details of a wire-type extensometer.

components of the extensometer system. This operation must be conducted precisely according to the manufacturer's instructions. When the readout components are reset, a reading before and just after resetting is taken to provide the constant to be added to or subtracted from future readings to determine total movement since installation.

   g. Data Processing and Review.—Data from extensometer observations are entered on a field data form as shown on figure 4-8, and transmitted to the Denver Office. Data processing and review are then conducted as outlined in section 1.25, and extensometer plots are prepared as shown on figure 4-9.

   4.5 Inclinometers (Deflectometers) Normal Installation.—a. *Usage.*—The terms "inclinometers" and "deflectometers" are used interchangeably regarding Bureau concrete dams. This manual uses "inclinometer" when referring to this type of measuring device. Inclinometers are used to measure lateral movements in abutments and foundations of concrete dams.

56

Inclinometer installations may consist of the "normal" type, where a casing is installed and a movable inclinometer probe is used to take readings; or a "fixed-type", where the inclinometer is fixed at a certain location in the casing.

  b. *Advantages and Limitations.*—Inclinometer installations have proven to give reliable and accurate data over an extended period provided proper installation, maintenance, and monitoring procedures are used. The only major limitation using this type of instrumentation is that a significant degree of specialized training is required for the personnel reading the devices. Such personnel must be trained in the proper care and operation of the equipment and should possess the ability to recognize erroneous or anomalous field data.

BRANTLEY DAM, PECOS RIVER, NEW MEXICO

Instrument: EX45R, Block 2, Station 87+86.5, 5.8' U/S

As-Built Downhole Anchor Locations from the Gallery Floor (Elevation 3205)

Anchor No.	Distance (Ft)
1	16.8
2	33.1
3	40.5
4	71.2
5	80.2
6	145.7
7	190.5
8	203.8

Instrument Readings

Date	Anchor No.	Mechanical Reading (In.)	Manual Reading (In.)	Date	Anchor No.	Mechanical Reading (In.)	Manual Reading (In.)
05/01/87	1	2.497	---	05/08/87	1	2.496	---
TEMP	2	2.465	---	TEMP	2	2.465	---
64° HI	3	2.471	---	62° HI	3	2.471	---
62° LOW	4	2.349	---	62° LOW	4	2.349	---
64° PRESENT	5	2.420	---	62° PRESENT	5	2.419	---
	6	2.408	---		6	2.408	---
	7	2.414	---		7	2.414	---
	8	2.393	---		8	2.391	---
05/14/87	1	2.497	---	05/22/87	1	2.496	---
TEMP	2	2.466	---	TEMP	2	2.466	---
64° HI	3	2.471	---	66° HI	3	2.471	---
64° LOW	4	2.350	---	64° LOW	4	2.349	---
64° PRESENT	5	2.420	---	66° PRESENT	5	2.420	---
	6	2.409	---		6	2.408	---
	7	2.417	---		7	2.414	---
	8	2.390	---		8	2.389	---

Figure 4-8.—Field data form for an extensometer.

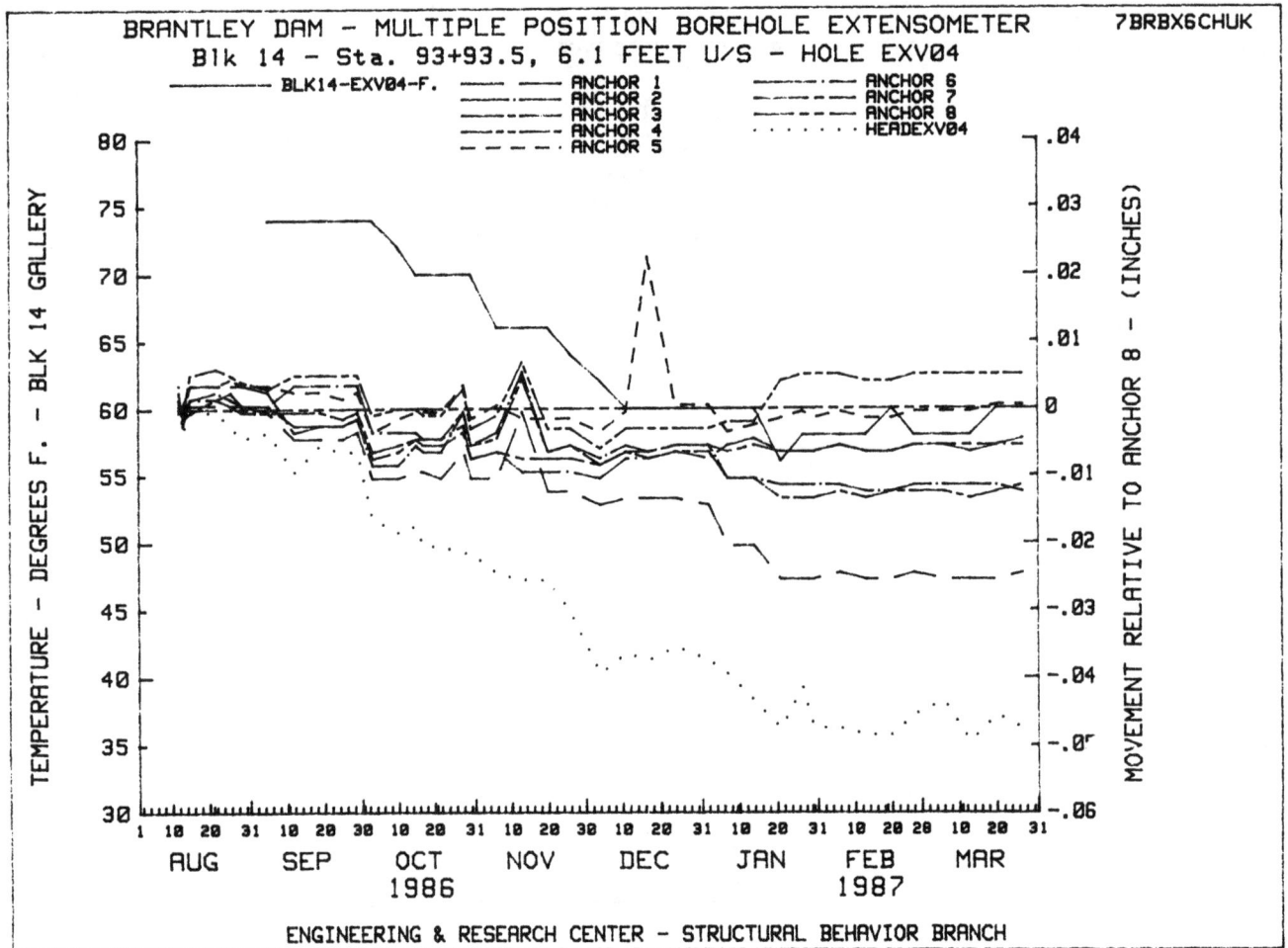

Figure 4-9.—Data plots on an extensometer.

c. *Description of Devices.*—Inclinometer systems may consist of special casings, casing couplings, inclinometer probe, cable, readout unit, data transmittal system, backfill material and, in some instances, a surface protective pipe with a locking cap.

(1) *Casing.*—Epoxy-coated aluminum, anodized aluminum, and plastic casings are available in various diameters and lengths. Selection of the proper size must include consideration of drill hole size (in a drillhole installation) and the intended use of the inclinometer. The inclinometer casing installation is the most critical factor in the entire inclinometer system. For example, if casing is to be used to monitor a landslide in an abutment that is actively moving, installation of small diameter casings is not practicable. When movement occurs, the smaller the casing diameter, the quicker the casing will close, and the hole, along with the possibility of additional data, will be lost. The drill hole should have a diameter as large as practical to enable a better grouting effort around the outside of the casing. For example, when installing the largest casing size, which is 3.38-inch (86-mm) outside diameter, a minimum 6-inch (152-mm) drill hole should be provided. Most types of casing are available in 5- or 10-foot (1.5- or 3.0-m) lengths.

Either plastic or aluminum casing with standard butt couplings may be used in drill hole installations. Plastic casing is not desirable for some installations principally because of possible warping due to extended exposure to high temperatures prior to installation. The Bureau has noted better overall results using aluminum casing rather than plastic casing in installations to date.

(2) *Casing Couplings.*—Two common types of coupling are available. For use in embankments, slip-type aluminum couplings of 12- to 24-inch (305- to 610-mm) length are available. Because this type of installation may also be used for settlement observations, the length of coupling should be selected so that sufficient length is available for anticipated settlement. In addition, larger diameter casings normally have longer couplings.

A standard 6-inch (152-mm) long coupling is used for butt joint drill hole installations; the coupling is attached to the casing by pop rivets at 90° points around the casing. The grooves in the casing must line up when the couplings are attached. Couplings are manufactured by the same companies that manufacture casing.

(3) *Inclinometer Probe.*—The inclinometer probe is a device that is capable of measuring angles of inclination from vertical in two planes oriented at 90° to each other. Some servoaccelerometers used to measure inclination angles have the capability of measuring up to 90° angles, while the more commonly used devices measure only up to 30° angles. Spring-loaded wheels maintain the probe within the grooves in the casing. It is important that the cable be of sufficient length to perform measurements in the deepest casing on the project. Markings on the cable are at 1-foot (0.3-m) intervals to enable accurate positioning of the probe for each reading.

(4) *Readout Unit and Data Transmittal System.*—Early models of readout units visually displayed the reading of one accelerometer and, by switching the selector switch, the reading of the second accelerometer was displayed. Both readings were recorded on the data sheet along with the depth measurement. The more recently developed cassette recorder unit is a self-contained, portable, waterproof unit with a rechargeable battery. Inclinometer data are visually displayed to the operator and simultaneously recorded on the cassette unit. The date, inclinometer number, and depth of reading are also recorded.

When the cassette recorder is used, the data transmittal system consists of a data selector switch and interfacing leads with the cassette recorder and a CRT terminal. An additional lead from the selector switch interfaces with a telephone modem to transmit the data to the Bureau's centralized computer system in Denver. All existing equipment operates at the 300-baud rate.

(5) *Surface Protective Pipe.*—A 1-foot (0.3-m) diameter, schedule 40, steel pipe with a thick steel lid is required to provide protection and security for the top of the casing in abutment installations. The surface protective pipe should extend into the abutment a minimum of 3 feet (0.9 m) or below the maximum anticipated frost depth. The lid is attached by a hinge on one side and has a hasp and eye on the other side.

(6) *Backfill.*—Inclinometer casings installed in drill holes are typically backfilled with a pumped sand-cement grout.

d. *Installation Procedures.*—Inclinometer casings must provide reliable orientation for the inclinometer probe; therefore, proper installation procedures must be observed. The locations for the installations are selected because of subsurface conditions, topography, and depth of any anticipated movements.

The casing may be placed in almost any stable drill hole that is sufficiently large enough to contain the casing and backfill. Some problems have been experienced with corrosion of aluminum casing when the casing has been exposed to alkaline soil, corrosive ground water, or certain grouts, especially in the presence of dissimilar metals. Coating the aluminum casing with epoxy has helped minimize this problem, but epoxy can be scratched or removed during installation. Flushing the casing with water after grouting has also been found to aid in minimizing corrosion. Anodized aluminum casing has been found to perform satisfactorily in most installations. To reduce drilling costs, casing is sometimes installed in conjunction with subsurface test boring and sampling programs.

Inclinometer casing is usually obtained in 5- or 10-foot (1.5- or 3.0-m) lengths, with the 10-foot-length being preferable. While drilling and sampling is being completed, the casing sections may be preassembled in 10-, 20-, or 30-foot (3.0-, 6.1-, or 9.1-m) segments to expedite the downhole assembly. The casing sections are joined with couplings, and sealed to prevent intrusion of grout or backfill materials. Depending on the application and type of casing, solvent cements, tape, or other sealers may be used. For plastic casing, a special alignment key is inserted into the grooves of each section to aid in alignment. For added strength and to aid in preventing twisting and groove misalignment, pop-rivets are used on both plastic and aluminum casing to reinforce the couplings.

One method of grouting involves the use of one-way valves that are fitted to the bottom section of the casing to facilitate grouting through the casing using drill rods. Once the casing is in place, hollow drill rods are lowered into the casing and attached to the nipple of the one-way grouting valve. Problems associated with this method of grouting are:

•As drill rods are being lowered, they may cut into the casing sidewall and damage the grooves.
•After grouting is completed, a large volume of water must be pumped down the drill rods to remove any remaining grout from the rods. If this is not done, the grout will spray the inside of the casing and adhere to it, which will create roughness or plugging of the grooves.

During insertion of the casing sections, clamps should be used at the top of the hole to prevent the lower casing sections from falling to the bottom of the hole. For very deep installations, a cable may be attached to the bottom of the casing to relieve the strain on the casing couplings. During the lowering process, the casing grooves should be kept aligned as near to final alignment configuration as possible so that spiraling caused by later turning of the casing will be minimized. Extreme care must be taken at all times to prevent damage to or misalignment of the casing.

The selection of backfill material, be it grout, sand, or pea gravel, depends primarily on the soil, rock, ground-water conditions, and annular space available around the casing. Grouting is generally preferred but may not always be feasible in the case of very pervious geological materials.

Grout usually consists of a cement-water mixture, and is pumped through a small diameter plastic pipe attached to the outside of the casing during installation. As previously discussed, grout may also be pumped through hollow drill rods and the bottom one-way valve until the annular space around the casing is completely filled. For very deep installations, grouting may be necessary at several levels. If grout is used with plastic casing, the heat of hydration of the grout or excess grout pressure can deform the plastic. This problem can be alleviated by maintaining the casing full of water until after the grout has set.

In many instances, the casing will need to be filled with water to overcome bouyancy in water- or mud-filled holes. Water inside the casing also helps to protect against casing collapse during grouting. The casing grooves may become misaligned during installation and grouting. If this occurs and the casing is still free to rotate, the casing may be given a final orientation adjustment so that one set of grooves is facing the direction of anticipated movement. This may be accomplished by slightly raising the casing, gently turning it, and then lowering it back to final elevation.

The top of the casing is usually capped with a plug to prevent intrusion of debris. In most cases, a steel protective pipe is installed around the top of the casing to discourage vandalism. The protective pipe is also fitted with a lid that may be locked in place. Typical inclinometer installations are shown on figures 4-10 and 4-11.

e. *Monitoring Procedures.*—The inclinometer installations are monitored at prescribed intervals. Inclinometer devices consist of a probe, cable, and digital readout unit. The probe is designed to fit any of the three standard casing sizes, which are 1⅞, 2¾, and 3⅜ inches (48, 70, and 86 mm). While setting up this instrument to take a set of readings, the guide wheels should be checked for bearing freeplay and tightened if necessary. A small amount of lubricant should be applied to each wheel bearing assembly. The cable connectors should be protected with caps to prevent damage or contamination to the connector contacts. An O-ring provides a watertight seal between the cable and probe; however, some probes are permanently connected to the cable. For the latter case, especially with lighter cables, the cable should be checked for damage near the connection.

The power to the readout unit is turned on and the system checked by holding the bottom of the probe stationary and moving the top along the measurement axis. The readout unit should display values with polarity corresponding to the given tilt angles. A calibration stand may be used if desired.

A pulley assembly is attached to the top of the casing and, usually, the probe must be installed in the casing before the pulley wheel can be fitted to the assembly. It is important that the wheels of the probe are properly seated in the premarked casing grooves before installing the pulley. The probe is carefully lowered to the bottom of the casing avoiding severe shocks to the instrument when nearing the bottom. On the Slope Indicator type of pulley assembly, a jam cleat is located at the top of the pulley assembly that is used for securing the cable and to provide a convenient depth index during the data collection process. The distance from the casing top to the jam cleat is 1 foot (0.3 m). The depth is controlled by pulling the cable through the cleat until the end of the foot mark has reached the rear of the jam cleat block. After the probe reaches the bottom of the casing, the probe is raised to the nearest foot mark, which is the distance from the top of the casing to the probe as measured by the cable markings.

When the inclinometer casing is initially surveyed, it is important that one groove be clearly marked and that the reference wheel be aligned with this mark for the initial set of readings. This is very important because the orientation of the readings is governed by the direction of this wheel. For every subsequent set of readings in this casing, the same wheel should be placed in the same groove used for the initial set of readings. A consistent system is highly desirable, and accurate records should be kept on the reference system used on a particular project.

A set of inclinometer measurements includes readings in opposite grooves (180° apart) at each depth. The sum of the readings in opposite directions at any given depth should be approximately constant for all

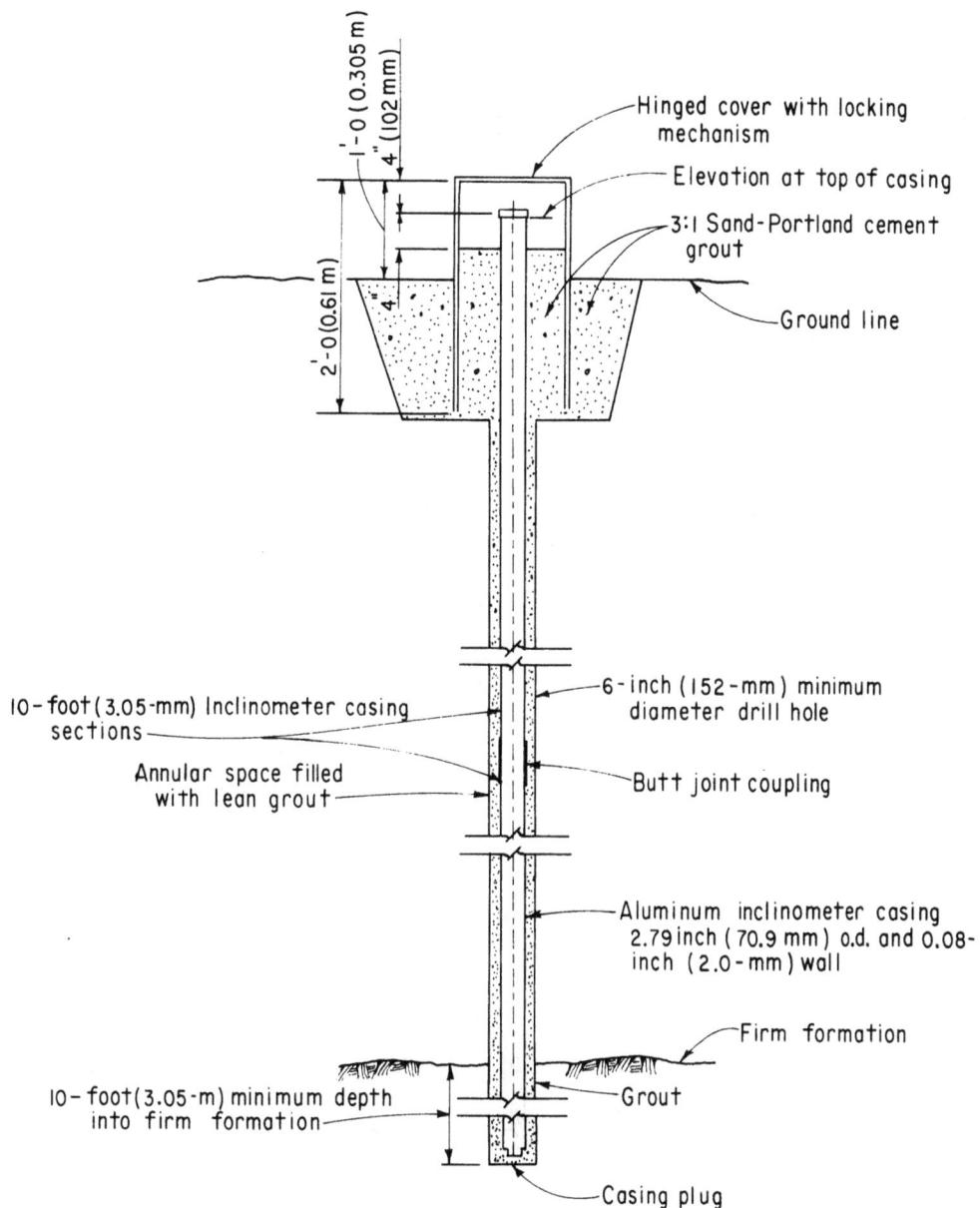

Figure 4-10.—Typical foundation and abutment inclinometer installation. 71–D–60.

measurements. Comparison of these sums will indicate any large errors that could result from mistaken transcriptions, faulty equipment, or improper technique.

When a cassette recorder is not being used, the data are entered manually onto the field data form in such a format as to facilitate quick conversion to computer coding. Because all inclinometer readings are referenced to an original set of measurements, extreme care must be taken to obtain an initial set of observations that are accurate and reliable. Measurements of the original casing profile should be established by at least two sets of data. If any set of readings deviates from the previous or anticipated pattern by a significant amount, the inclinometer should be checked and readings repeated.

Careful attention must be given to depth measurements on the cable to realize the potential accuracy of the inclinometer system. The wheels of the inclinometer probe provide measuring points between which the inclination of the instrument is measured. To achieve optimum accuracy, the distance between each reading interval should equal the distance between the upper and lower wheels of the probe. Taking two or more data sets each time also serves to improve accuracy.

Despite precision manufacturing techniques used by most casing suppliers, the grooves in any section of casing may be slightly spiraled. During installation, the casing may become even more twisted so that at some

61

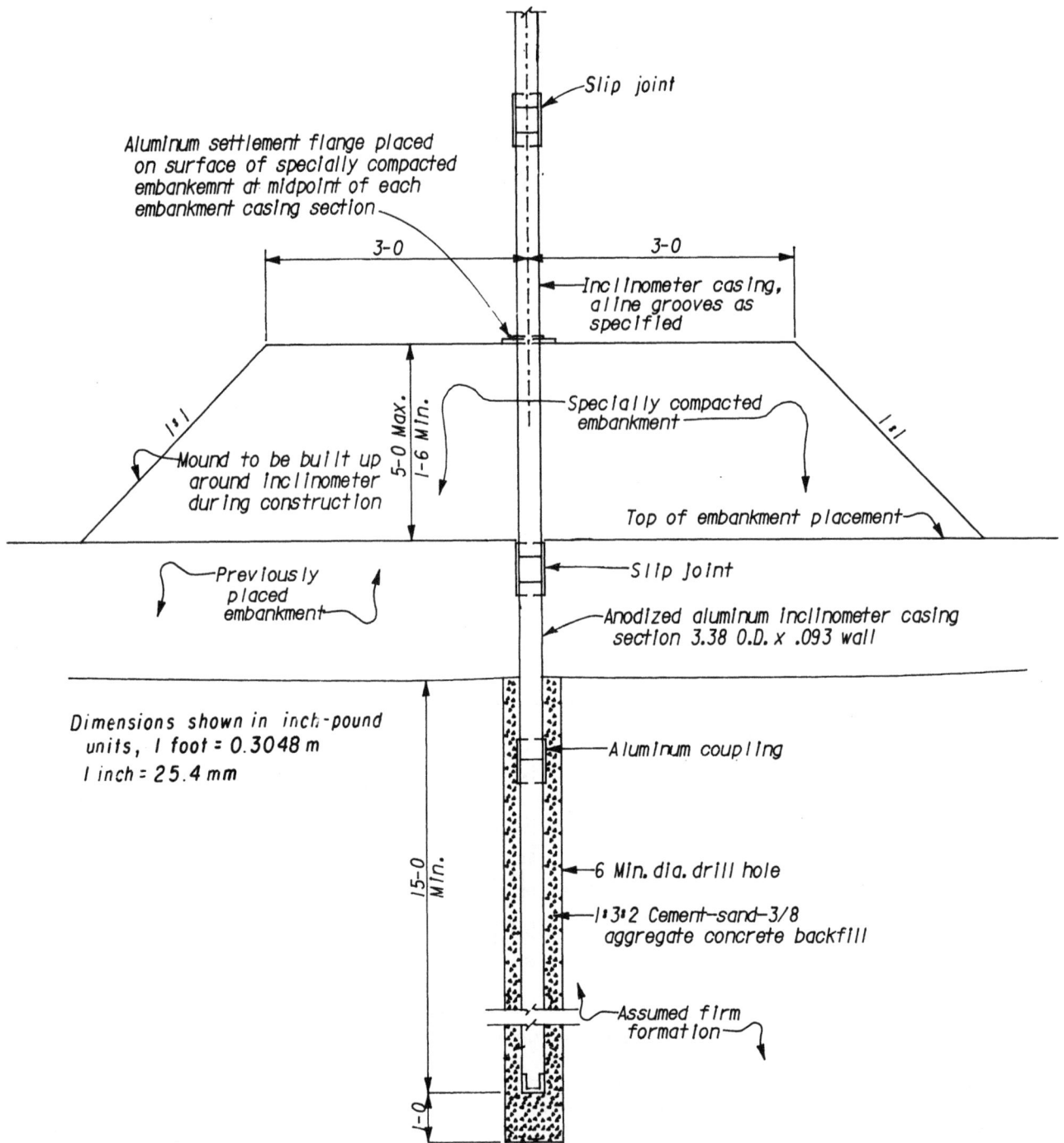

Figure 4-11.—Typical embankment inclinometer installation. 963-D-127.

depth, the casing grooves will not have the same orientation as at ground level. Because significant errors in the assumed direction of movement may result, casing for deep inclinometer installations should be selected with great care and checked for spiral before installation. Spiral measuring probes are available from the manufacturers but have, in the past, had a somewhat unreliable performance record. Undetected spiraling may not contribute to an error in magnitude of deformation but will likely prevent determination of the true direction of movement of the casing.

Automatic data recording and processing are useful in reducing the time and labor involved in computation of the data. However, automatic data recording contributes to the complexity of the measurement operations and can also introduce an additional and significant source of error. With manual data recording, a technician can scan the data for errors in the check sums and make corrections, or reread the casing. Most data recording systems do not allow this advantage; the data must be scanned for errors after being printed out in the office and before computer processing. One data recording system presently allows a review of the data as it is being taken so that corrections can be made as measurements are accumulated. This system also allows for field reduction of the data and comparison of the data to initial data files. Use of the same observers and instruments for all measurements on a particular project is strongly recommended. All data except that recorded and transmitted automatically are entered on a field data form as shown on figure 4-12.

f. *Maintenance.*—A regular maintenance program is essential to the proper conduct of inclinometer monitoring. Repair or replacement of an inclinometer probe or readout unit due to misuse or failure to properly maintain the devices can be expensive and result in loss of data due to downtime.

Upon completion of field work, the unit should be checked for superficial damage. Wheel fixtures and bearings should be checked for tightness, cleaned, and oiled. If worn, these parts should be replaced as soon as possible. Metal surfaces should be cleaned and protected with a coat of light oil. The cable should be inspected for damage to the outer coating material and to the depth markers. If depth markers are not vulcanized to the cable, they may move and result in erroneous depth readings.

Cable connectors should remain capped when not in use and thoroughly cleaned after use. Most readout units are not completely weatherproof and should be protected from moisture. The introduction of a few drops of water into the electronic circuitry can cause a galvanometer to read incorrectly or could induce a drift on a digital voltage display. If readings are being made rapidly or if an automatic recorder is used, this drift may not be detected.

Equipment should be protected from shock or excessive vibrations during transportation, and storage of this sensitive electronic equipment in a closed vehicle during the summer or in an unheated area in winter should be avoided.

When using an automatic data recorder, additional periodic maintenance is required. Battery maintenance guidelines require the batteries to be completely discharged after a certain number of charge/discharge cycles. Nickel-cadmium batteries retain a form of memory capacity such that if they are discharged by 10 percent of their capacity and then recharged to full capacity repeatedly, they may soon deliver only about 10 percent of their potential capacity. Complete discharge is defined as 5.0 volts. The minimum voltage of the internal batteries that will safely operate the unit should be carefully monitored because a lower voltage could result in lost data. The tape cassettes used with the recorder should be protected against dust, dirt, and temperature extremes; and should be cleaned periodically as prescribed by the instrument manufacturer.

Periodic maintenance should also be conducted on the cable reel if one is used. The rotating electrical contacts on the cable reel require periodic cleaning; disassembly of the reel is required to service or replace these contacts. All system components should be stored in a dry, moderate temperature environment when not in use.

g. *Data Processing and Review.*—Data from the periodic monitoring are transmitted to the Denver Office and processed as described in section 1.25. A data plot illustrating lateral movement is shown on figure 4-13.

4.6 Inclinometers—Fixed Position.—a. *Usage.*—Fixed-position inclinometers are used where, for various reasons, it would be difficult or impossible to read a normal inclinometer installation.

b. *Advantages and Limitations.*—The inplace or "fixed" inclinometer is a solid-state borehole inclinometer used for measurement of progressive changes in the angle of inclination at set locations within an inclinometer casing. The fixed position inclinometer has an advantage over conventional inclinometers in that it can be installed in locations that later become inaccessible to instrumentation personnel; i.e., the upstream face of a dam under the reservoir or any location not readily accessible. The system can then be remotely read. The fixed inclinometer has two principal limitations: (1) only selected segments along the profile of the hole are measured rather than the entire hole depth, and (2) less economical than other inclinometers.

c. *Description of Devices.*—The fixed inclinometer system consists of an inplace sensor, interconnecting electrical cable, junction box, permanently installed inclinometer casing, remote terminal box, and readout equipment. The casing and readout equipment were previously described in section 4.5.

(1) *Sensor.*—The inplace sensor is very similar to the standard inclinometer probe described in section 4.5. An important exception is that the inplace sensor monitors inclination over only a given gauge length

7-1884A (4-74)
Water and Power

BOREHOLE DEFLECTOMETER FIELD DATA
(BIAXIAL INCLINOMETER)

PROJECT	FEATURE
FRY-ARK	PUEBLO DAM

INSTR. NO.	CAL FACTOR	READ	CHKD.	SET NO	SHEET
TPC-100		MUSGROVE	CIRVLI	069	1 OF 1

HOLE NO.	STATION	OFFSET	DATE	TOTAL READINGS
05	DF-5 89+99.8	290-DS	06-07-83	029

DEPTH	A COMPONENT	B COMPONENT	A COMPONENT	B COMPONENT	Direction of Spring-Loaded Wheel
5.	-0.099	-0.314	-0.089	+0.351	NORTH GROVE
7.	.080	.294	.110	.345	FIRST, THEN
9.	.089	.293	.112	.341	2 SOUTH GROVE
11.	.099	.297	.102	.345	3
13.	.099	.361	.103	.395	4 LEVEL OF
15.	.015	.379	.165	.419	5 H₂O IN
17.	.052	.361	.149	.419	6 DEFLECTOMETER
19.	.054	.371	.135	.418	7 TUBE 57±
21.	.060	.354	-0.147	.395	8
23.	.209	.339	+0.014	.402	9
25.	.337	.380	.137	.419	10
27.	.329	.381	.132	.435	11
29.	.314	.381	.132	.430	12
31.	.329	.380	.139	.422	13
33.	.314	.350	.127	.357	14
35.	.322	.395	.129	.462	15
37.	.254	.615	.392	.667	16
39.	.327	.252	.215	.287	17
41.	.339	.215	.159	.249	18
43.	.359	.190	.185	.263	19
45.	.352	.235	.149	.263	20
47.	.317	.217	.123	.249	21
49.	.314	.197	.120	.224	22
51.	.281	.197	.065	.249	23
53.	.240	.265	+0.022	.334	24
55.	.173	.282	-0.023	.290	25
57.	.159	.254	-0.047	.289	26
59.	.154	.255	-0.049	.299	27
60.	-0.154	-0.261	-0.047	+0.292	28
					29
					30
					31
					32
					33
					34
					35

Figure 4-12.—Field data form for an inclinometer.

FRY-ARK
PUEBLO DAM
HOLE NO D-11
INCLINOMETER READINGS DATED 1-07-83
FROM SET OF READINGS, SET NO. 71

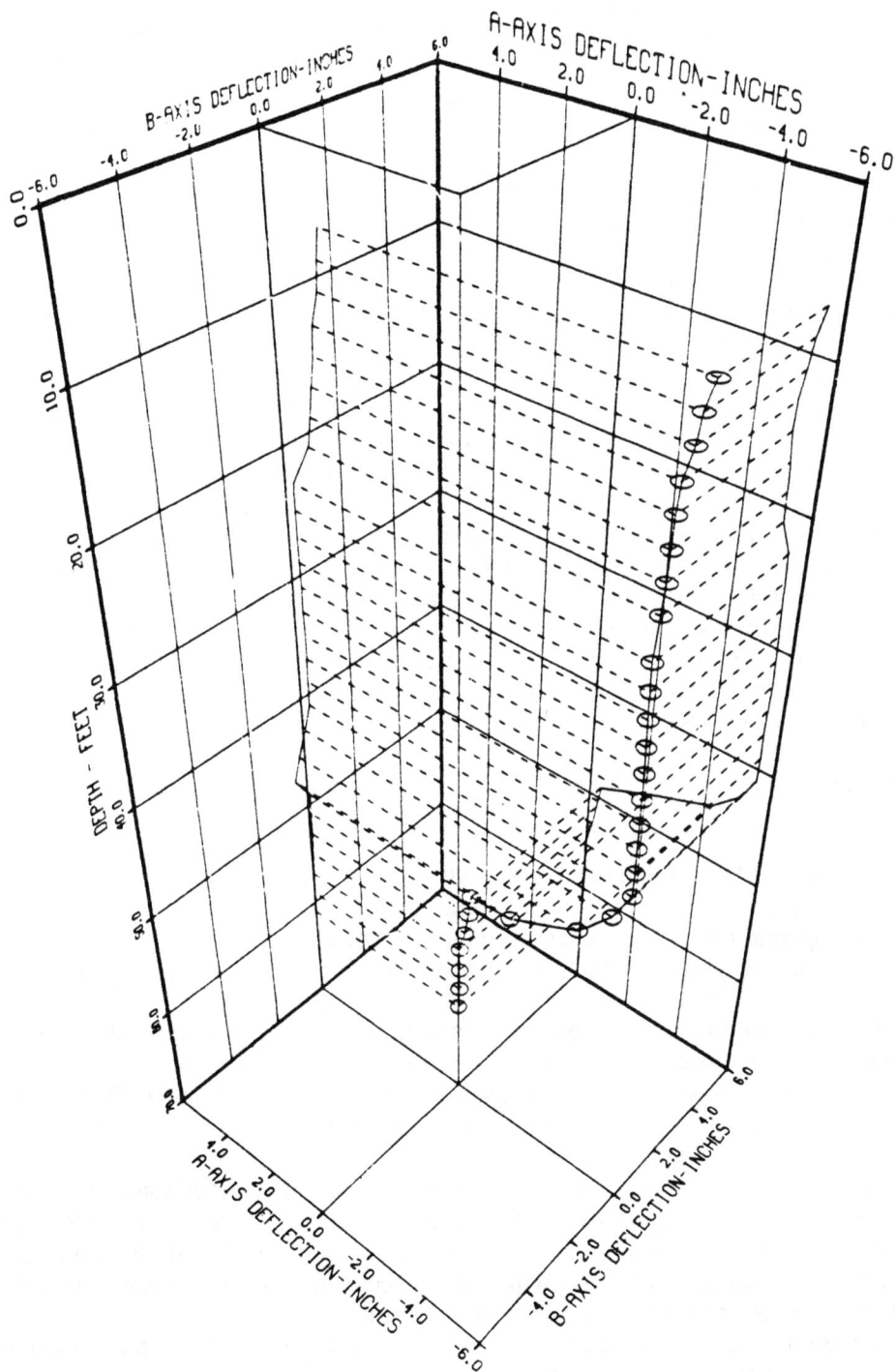

Figure 4-13.—Data plots on an inclinometer.

at a certain location in a hole. The sensor contains two servo-accelerometers mounted with the sensitive axes 90° apart to measure the angle of inclination of the longitudinal axis of the sensor in two orthogonal planes. The sensor is installed nearly vertical in the inclinometer casing and is supported laterally in the casing by guide wheels. The sensor has a range of $\pm 30°$ with a resolution of 0.001 inch (0.025 mm) for a 10-foot (3-m) gauge length. The gauge length is the vertical distance between the universal joints at the centerline of the guide wheels of the sensor and the end of the gauge length tubing.

(2) *Electrical Cable.*—The sensor has an electrical cable that terminates at a junction box at the top of the casing or at a remote location. This cable has six stranded No. 20 AWG copper conductors, a single braided shield overall, and a polyethylene jacket.

(3) *Junction Box.*—The junction box can be installed at the top of the hole or at a remote distance from the hole, and is able to read up to 16 sensors installed in the hole. The junction box is intended for manual readout of the sensors. A multipin connector is provided on the outside of the box for quick connection to the readout equipment. Fixed inclinometers can be fitted with alarm systems to give warning in case of excessive movement. If alarms are installed, an excitation and alarm unit is attached at the junction box. Up to 30 sensors can be used to trigger a single alarm unit.

d. Installing Sensors.—Prior to installation, the inclinometer casing should be surveyed as described in section 4.5. These data should be reviewed with respect to the verticality of the hole so that full scale limits will likely not be exceeded by the sensor. Also, sharp bends at or above the depth of sensor installation, which would prevent sensor installation, should be checked. The casing should be permanently marked on the groove that is used to determine the "A+" axis of the casing. The fixed wheel of the sensor will be placed in this groove during installation.

The support tubing and sensors are arranged in order of installation, and the sensors are connected to their gauge length tubes. All sensors are then tested for proper A and B axis polarity and stability.

The sensors are installed by lowering both sensors and tubing into the casing and aligning the A+ polarity of the sensor to the A+ groove on the casing. The sensor and connection tubing are lowered using a steel retrieval cable attached to the bottom sensor. This cable must be attached to the tubes during installation to avoid tangling with electrical cables. As the sensor's second set of wheels is placed in the casing, a special alignment jig is used because of the universal joint on these wheels. Once sensors and tubes are all in place, the tubes are attached to the casing by a special cross which fits inside the casing. The sensor should then be checked for operation. The sensor can be removed using the retrieval cable if a complete profile by a movable inclinometer probe is required, see figure 4-14.

e. Monitoring Procedures.—Sensors from many locations or many depths can be read at one central location. The cost of the cable is a factor in placing the reading location as close as possible to the sensors. The sensors are read by connecting the readout device to the junction box, then completing the following steps:

1. Turn power switch "ON" and allow readout 10 minutes for warmup.
2. Turn power switch to "Battery Test," the digital display should indicate 6.0 volts for a fully charged battery. Do not operate readout if voltage is below 5.5 volts.
3. Set the junction box selector switch to "Sensor No. 1," allowing at least 1 minute for sensor to stabilize.
4. Record magnitude and polarity of reading displayed. Both A and B components should be recorded. Some readout devices require switching from A component to B component.
5. Set junction box selector switch to the next sensor and repeat steps 3 and 4. Whenever selector switch is set to a new sensor, 1 minute should be allowed for sensor to stabilize.

f. Maintenance.—The periodic maintenance required by the inplace inclinometers is accomplished by properly maintaining the readout device (sec. 4.5) and junction box. To ensure good electrical contact, periodically wipe contacts with a clean dauber or cloth previously dipped in alcohol. Before disconnecting cable from the junction box after readings, carefully wipe the connector to remove any water or dirt present. When disconnected, always keep the protective cap over the contacts.

g. Data Processing and Review.—The data readings obtained are in a form that usually can be used to check for any movement at the time the measurements are taken. Because the purpose of the inplace inclinometer is to monitor changes in the deflection of the inclinometer casing, it is necessary to summarize the

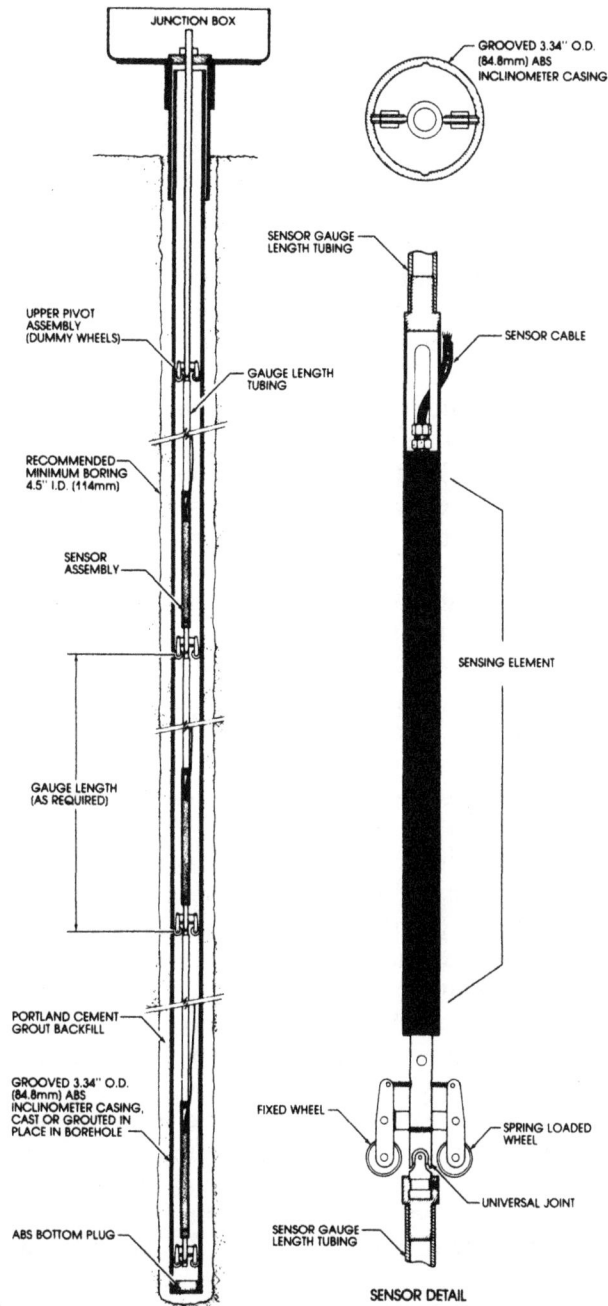

Figure 4-14.—A fixed position inclinometer installation.

data on data sheets to compare successive surveys. The field data can be tabulated on the data sheet by showing the reading and date taken. The data do not usually need to be converted to angles to be reviewed. The periodic change in deflection is determined by simply subtracting the initial reading from the new reading. The initial reading is used as a baseline reference for all future readings. If the inclination is required, the tilt angle for any single reading at any depth may be determined as follows:

$$R = 12\ L \sin\ \theta,\ \text{or} \tag{1}$$

$$\theta = \text{Arcsin}\ \left(\frac{R}{12L}\right) \tag{2}$$

67

where:

R = instrument reading,
L = gauge length in feet (meters), and
θ = tilt angle in degrees.

The data are then transmitted and processed as described in section 1.25.

4.7 Strain Meters.—a. *General*.—The majority of embedded strain meters used by the Bureau in the past have been of the type developed by R. W. Carlson [10]; however, the Bureau now usually uses the vibrating-wire type. The Carlson instruments use the electrical principles that changes in tension in an elastic wire cause changes in electrical resistance of the wire, and changes in temperature of the wire also causes changes in electrical resistance of the wire. Strain, joint, and stress meters all utilize these principles to measure strains, displacements, stresses, and temperature changes. In the internal resistance thermometer, temperature changes are measured by resistance changes in a copper wire.

b. *Advantages and Limitations*.—Strain meters have proven to be accurate and reasonably reliable for Bureau dams; however, they do require care in installation and in obtaining readings. Strain is only measured at the location of the strain meter; therefore, it is necessary to carefully plan the locations for the proposed strain meters to optimize the amount and applicability of the information obtained.

c. *Description of Devices*.—Strain meter installations consist of one or more strain meters, a "spider" frame for mulitple meter installations, cable, and a readout unit.

(1) *Strain Meter*.—The strain meter device consists of a small diameter cylinder with flanges on each end to anchor the ends of the meter to the surrounding concrete (fig. 4-15). A steel framework within the flexible brass cover has two porcelain spools attached to it. Two equal coils of 0.0025-inch (0.06-mm) diameter fine steel music wire are wound under a 100,000-lb/in² (689-MPa) tension between the spools. The meter is filled with oil to prevent corrosion, and a small amount of air to allow for expansion. The oil acts as a heat sink during readings as well as providing protection for the wire.

When the distance between the ends of the strain meter is increased, the outer or expansion coil elongates and increases in tension and resistance. At the same time, the inner or contraction coil decreases in tension and resistance as it shortens. Changes in the ratio of the resistance of the outer coil to the inner coil are used as a sensitive measure of length change in the strain meter. The ratio of the resistances of the two coils is nearly equal to unity at all times. A change in ratio of 0.01 percent usually indicates a strain change of about 4×10^{-6} inch (102×10^{-6} mm) per inch (mm) of strain meter length. Resistance ratio changes are affected (but only insignificantly) by simultaneous changes in temperature of the wires because temperature change affects both coils by a nearly equal amount.

Temperature is measured by determining the series resistance of the two coils. The total resistance is not materially affected by changes in resistance of the two coils due to length change because these changes are substantially equal and opposite. The series resistance of the two coils in a strain meter is about 70 ohms at 70 °F (21.1 °C) and changes about 1 ohm for each 9 °F (5 °C) change in temperature.

The indicated length change represents the actual length change, providing there is no change in temperature. If there is temperature change, the indicated length change must be corrected for thermal expansion or contraction of the meter frame. For a typical meter, a correction of 7.5×10^{-6} inch (190.5×10^{-6} mm) must be added to the indicated unit length change for each 1 °F (0.5 °C) rise in temperature, or subtracted for a decrease in temperature.

The range of the meters is typically about 0.005 inch (0.13 mm) per inch (mm) of strain, with the range in contraction from the neutral or initial position accounting for about two-thirds of the range and expansion for about one-third. Built-in stops are provided to prevent breakage of the contraction coil that could result from mishandling of the meter. Strain meters are delicate instruments and must be handled with care and protected against any type of damaging treatment.

(2) *Cable*.—The cable for a strain meter, not usually furnished with the meter, normally consists of waterproofed, four-conductor, color-coded cable. The green and white conductors are common to one end of both coils, the red conductor connects to the other end of the expansion coil, and the black conductor to the other end of the contraction coil. A four-conductor cable system is used to allow for correct temperature measurement without the effects of cable resistance and changes in cable resistance due to temperature. The conductor wires are usually of No. 16 or 18 size. If additional conductor cable must be added in the field to reach from the meter to the readout area, the lead cables must be properly spliced to avoid shorts due to moisture penetration through the cable covering at the splice. The cable resistance must then be determined for calibration purposes.

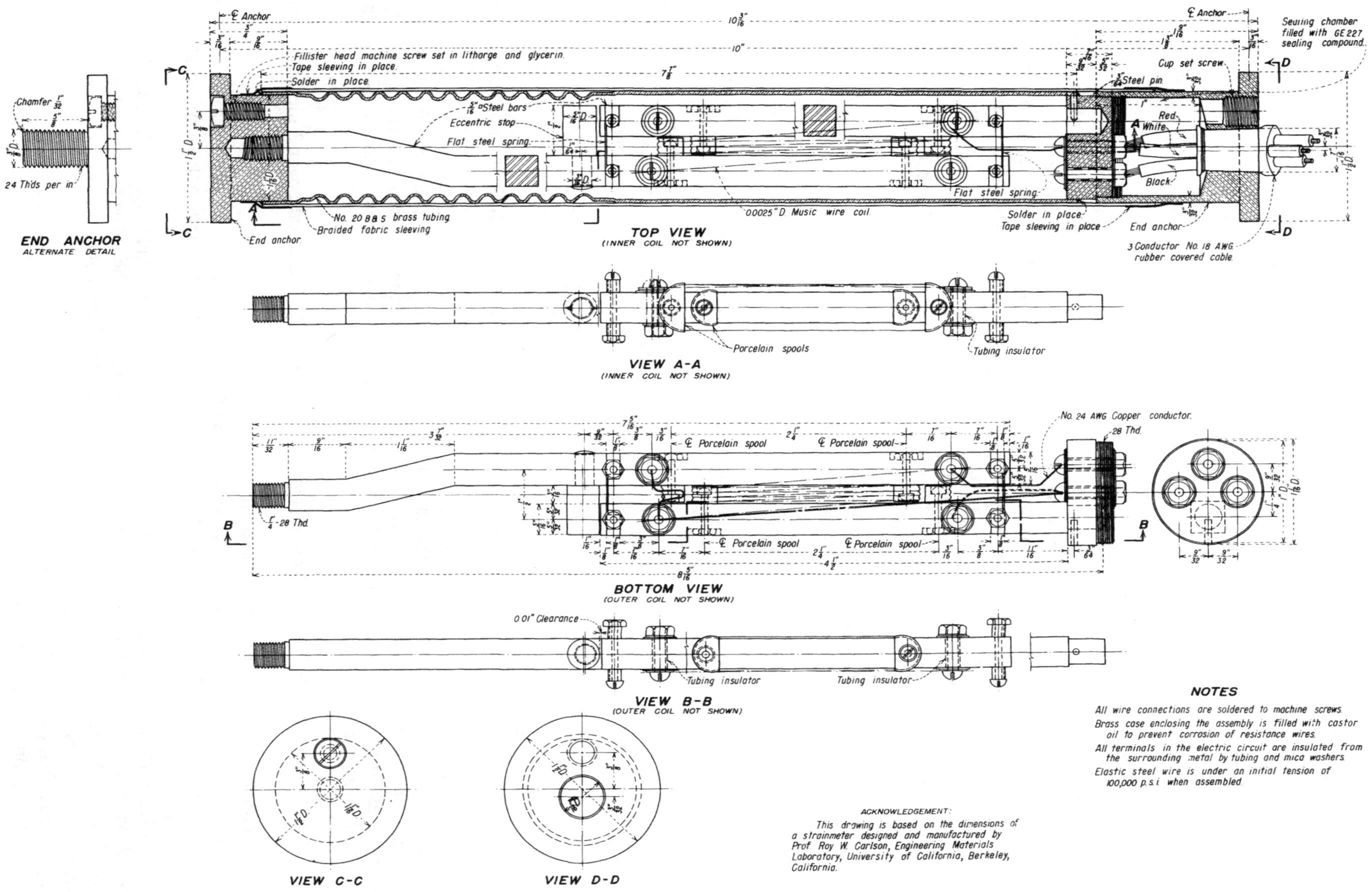

END ANCHOR
ALTERNATE DETAIL
Chamfer
24 Th'ds per in
Anchor
Fillister head machine screw set in litharge and glycerin.
Tape sleeving in place.
Solder in place.
Steel bars
Eccentric stop
Flat steel spring
No. 20 B & S brass tubing
Braided fabric sleeving
End anchor
0.0025" D. Music wire coil.
TOP VIEW
(INNER COIL NOT SHOWN)
Flat steel spring
Solder in place
Tape sleeving in place
End anchor
3 Conductor No. 18 AWG rubber covered cable.
Cup set screw.
Steel pin
Red
White
Black
Sealing chamber filled with GE 227 sealing compound.
Porcelain spools
Tubing insulator
VIEW A-A
(INNER COIL NOT SHOWN)
No. 24 AWG Copper conductor.
28 Thd.
Porcelain spool
-28 Thd.
BOTTOM VIEW
(OUTER COIL NOT SHOWN)
0.01" Clearance
Tubing insulator
VIEW B-B
(OUTER COIL NOT SHOWN)
VIEW C-C
VIEW D-D
ACKNOWLEDGEMENT:
This drawing is based on the dimensions of a strainmeter designed and manufactured by Prof. Roy W. Carlson, Engineering Materials Laboratory, University of California, Berkeley, California.
NOTES
All wire connections are soldered to machine screws.
Brass case enclosing the assembly is filled with castor oil to prevent corrosion of resistance wires.
All terminals in the electric circuit are insulated from the surrounding metal by tubing and mica washers.
Elastic steel wire is under an initial tension of 100,000 p.s.i. when assembled.

Figure 4-15.—Details of a typical strain meter. 40-D-4508.

(3) *Multiple Meter Frame.*—Figure 4-16 illustrates a multiple meter frame or "spider" used to position strain meters in clusters so that they will retain proper orientation during embedment in concrete. Eleven strain meters may be installed in this manner, oriented in the horizontal, vertical, and angular planes. In addition, a 12th strain meter is usually installed in a vertical orientation adjacent to the strain meter cluster. Figure 4-17 shows a horizontal stress meter, lower portion of photograph, being installed with a group of strain meters.

(4) *Readout Unit.*—The strain meters and their connecting cables are terminated at terminal boards as shown on figure 4-18. These boards are located at appropriate reading stations in the system of galleries within the dam. At each station, readings from the instruments may be obtained with a special Wheatstone bridge test set as shown on figure 4-19.

d. Installation Procedures.—(1) *Instrument Preparation.*—Upon receipt of a shipment of strain meters, they should be unpacked, inspected for damage, and checked for operability. The instruments should be checked for oil leaks, and then readings taken. The readings should be compared with calibration data furnished by the manufacturer, and the ratio reading should be very close to the neutral ratio also given in the calibration data. Instruments that show oil leaks, are unreadable, or give obviously unreasonable readings should be separated from the remainder of the shipment and immediately returned to the manufacturer.

Each meter's identification number should be carefully recorded, and the number put on any cable splices made so that the instrument identification is known at the final readout location. Any splices made should be in strict accordance with the manufacturer's recommendations. Obviously, if the instruments can be obtained and the proper cable length attached, the necessity for splices and their accompanying problems is avoided.

(2) *Calibration Corrections.*—Each instrument is individually calibrated by the manufacturer, who typically uses short cable leads. Adjustments to this calibration must be made using the actual lead lengths to be used when installed in the dam. The extra cable lead length will have a ballasting effect on the resistance ratio readings. The instrument characteristics furnished by the manufacturer, together with the calibration corrections, should be recorded in a permanent project file.

The calibration factor furnished by the manufacturer, strain change per unit change in resistance ratio, is corrected by multiplying the given factor by the ratio of the measured series resistance of the two instrument coils and the actual cable leads to the measured series resistance of the same two coils and the short cable leads. For this reason, careful measurement of the resistance of each cable conductor is imperative. The corrected calibration factor will always be greater than the initial factor because the long cable lead introduces additional resistance into the circuit.

An additional calibration check and possibly an additional correction may be necessary at the time the cable leads are connected to a readout terminal board, particularly if this operation involves cutting off significant lengths of cable.

(3) *Installation.*—The actual installation of strain meters in a dam is a difficult task that requires very precise placement and extreme care in instrument handling. A successful strain meter installation requires that each meter be aligned with reference to the structure axis or centerline as specified on the drawings, and that the concrete surrounding the meter be identical with the concrete used in the structure. Obviously, these goals cannot be completely achieved, but every effort should be made during placement operations to satisfy these conditions as far as practicable. Single strain meters are usually embedded near the top of a concrete lift in accordance with the following steps:

1. Dig into the area for the full depth of the instrument and discard all aggregate over 3 inches (76 mm) in size.
2. Backfill sufficiently to provide a bed for the instrument.
3. Make the hole for vertical or diagonal meters using an electric vibrator or by driving a 1.5-inch (38-mm) diameter steel pipe into the concrete, and then insert the strain meter into the hole. Horizontal meters are merely laid flat on the prepared bed.
4. Check angles, direction, and depth using a large protractor and a level held against the end flange.
5. Carefully vibrate or hand-puddle the concrete around the meters. Care should be taken to avoid excessive vibration that could disturb the meter alignment or break the small resistance wires.
6. Ensure that meter is still operable, replace if damaged.
7. Complete backfilling by hand up to grade.
8. Record all location data for each meter.

70

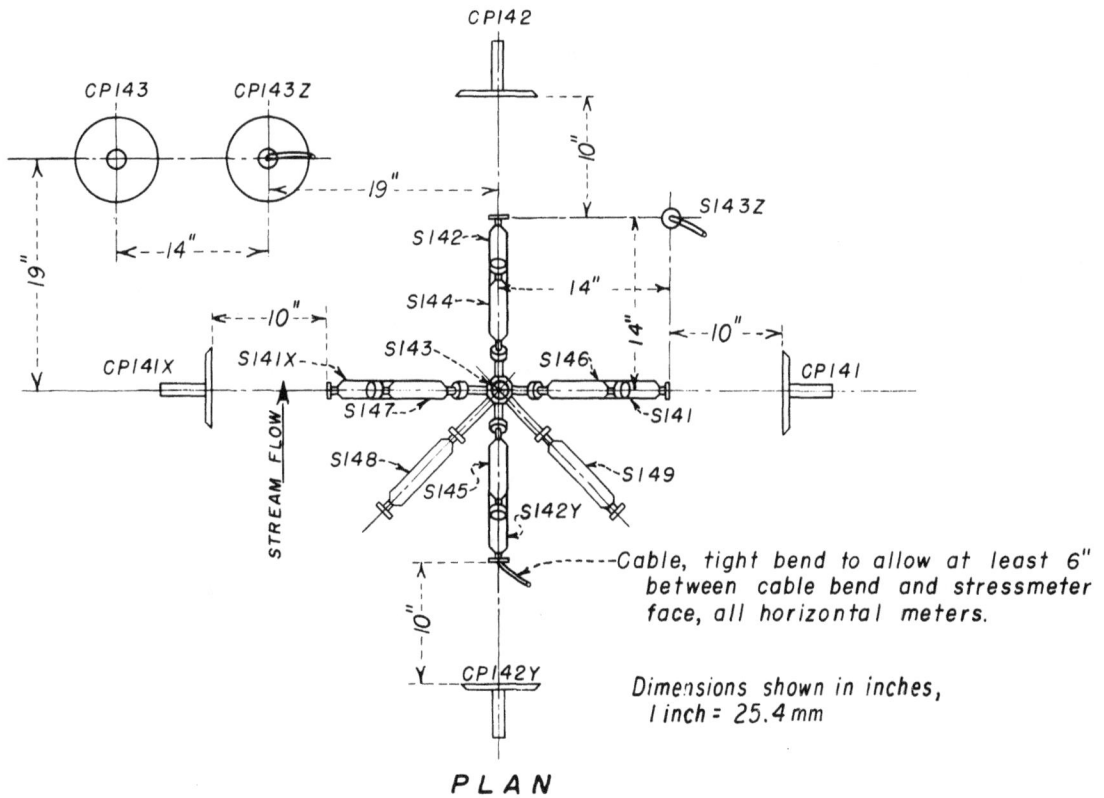

PLAN

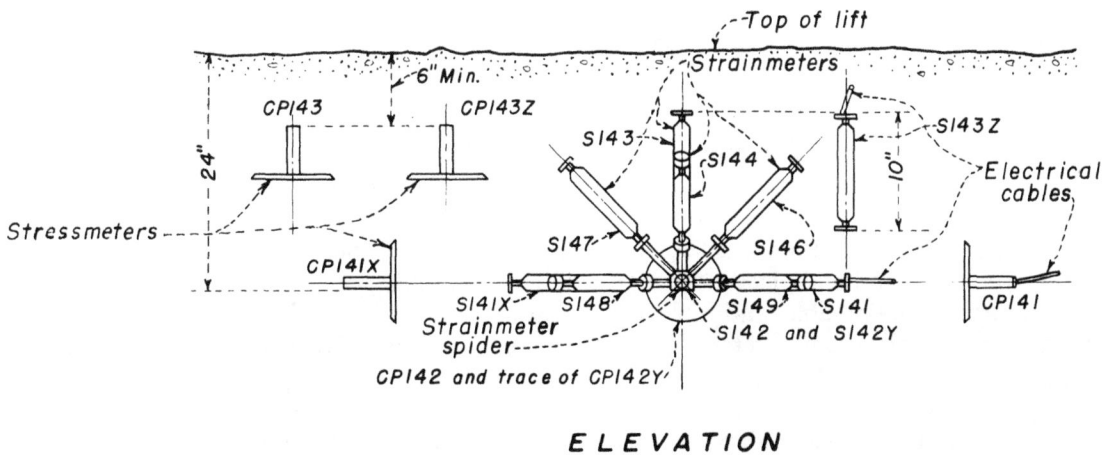

ELEVATION

Figure 4-16.—Details of a multiple strain meter frame. 557-D-2135.

Interior groups of meters require more elaborate preparations and facilities to assure proper installation in the limited time available before the concrete attains its initial set. A typical procedure for placing a group of meters is:

1. Place a temporary wooden blockout frame on top of the concrete lift when it has reached an elevation about 12 inches (305 mm) below final grade.
2. Finish off remainder of lift outside the frame.

Figure 4-17.—Horizontal stress meter (lower portion of photograph) being installed with a group of strain meters at Glen Canyon Dam. P-557-420-5879.

3. Explore the bed inside the frame and remove cobbles larger than 3 inches (76 mm) in size.
4. Set a previously prepared template using survey points, and place meters at their approximate locations.
5. Using a vibrator or a drivepipe, make holes in the concrete at the proper locations and inclinations.
6. Place meters in the prepared holes and check for proper location and inclination.
7. Cover each instrument with hand-placed concrete, and then carefully compact.
8. Ensure that all meters are operable, and replace if necessary.
9. After concrete has stiffened slightly, remove blockout frame, fill remainder of hole with concrete, and carefully compact.

Multiple meter installations using prefabricated "spider" frames simplify the installation procedure as follows:

1. Level a concrete bed 2 feet (0.6 m) below the ultimate elevation of the concrete lift.
2. Mark meter group location with a light board framework and finish up the lift around this location.
3. Group cables together on a nearby plywood platform and make final assembly of strain meters on the spider.
4. Inspect the bed and remove cobbles larger than 3 inches (76 mm).
5. Place spider and attached meters in position and check orientation and level.
6. Carefully place and compact concrete around meters by hand.
7. Protect area from disturbance until concrete sets.

72

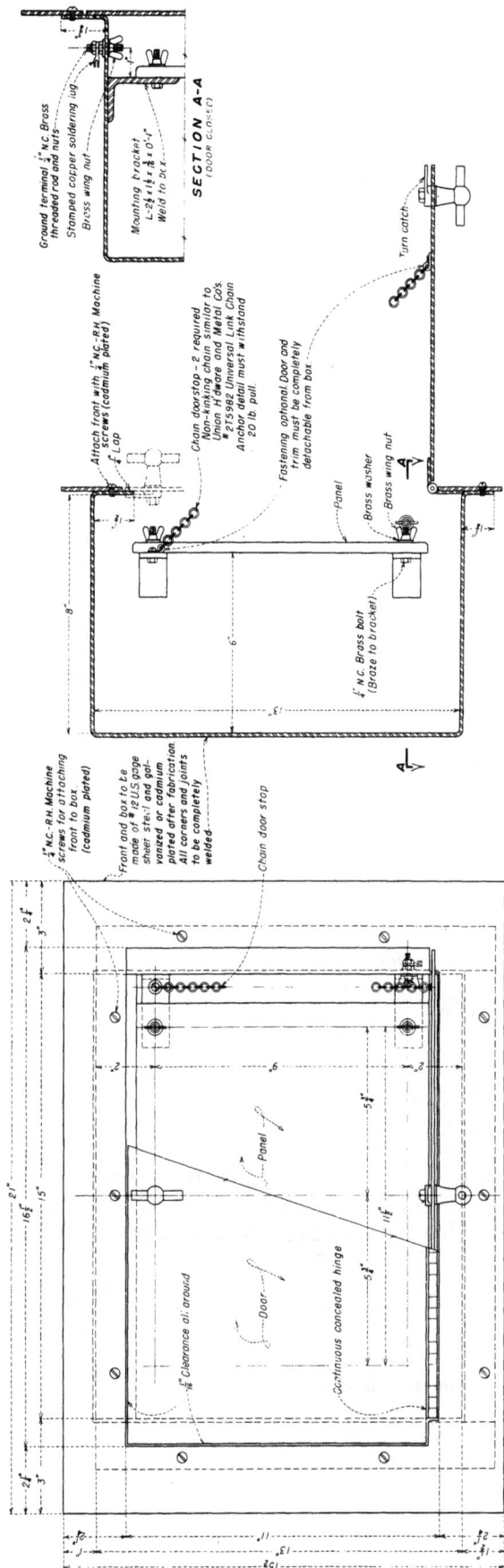

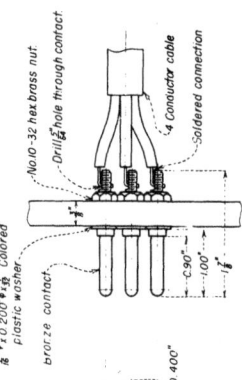

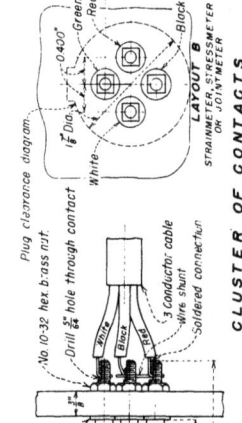

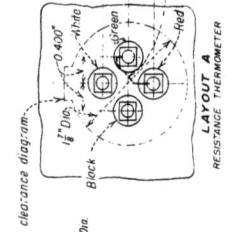

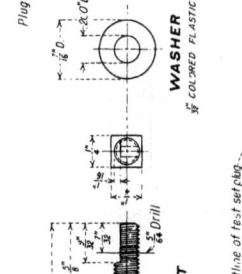

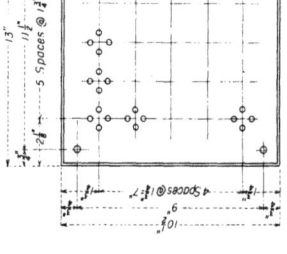

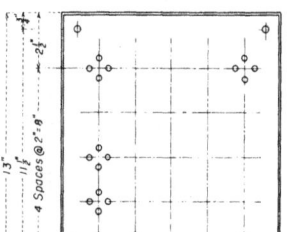

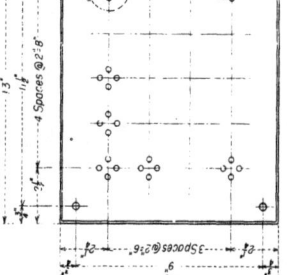

Figure 4-18.—Details of a strain meter terminal board. 40–D–4868.

73

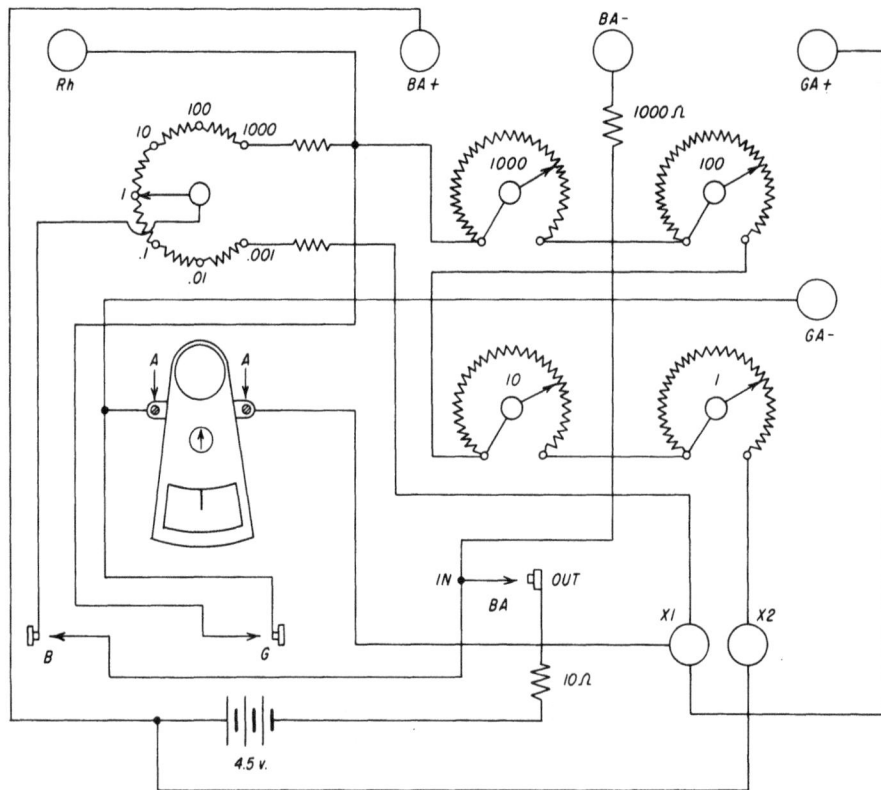

WIRING DIAGRAM

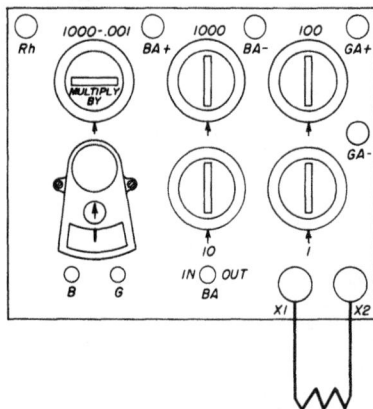

FOR MEASUREMENT OF RESISTANCE
(.001 - 9,999,000 OHMS)

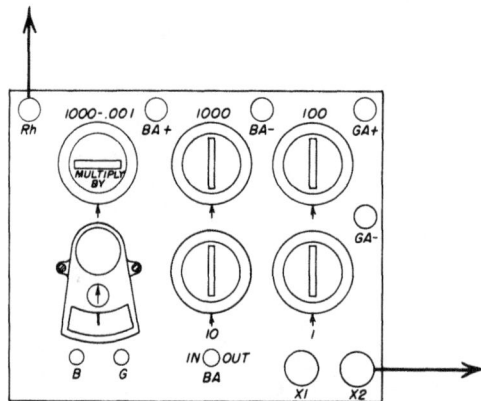

FOR USE AS A VARIABLE RESISTANCE
(1-1000 OHMS)

Figure 4-19.—Details of a Wheatstone bridge test set. 40-D-4954.

Surface or boundary strain meters are usually placed in groups of three to six at a distance of 3 inches (76 mm) to 3 feet (0.9 m) from a surface of the structure. Each meter is arranged parallel to the concrete face and in a vertical plane. Positioning of the meters at the required distance from the face and in the proper orientation can be attained by using special pipe brackets bolted or fastened to the top of the forms. Each bracket holds a length of 1.5-inch (38-mm) diameter pipe oriented at the proper distance and angle. When concrete is placed, each pipe forms a hole slightly larger than the meter diameter. A shallow hole is dug around each pipe to about 8 inches (203 mm) in depth. After concrete has stiffened slightly, the pipes are pulled out and strain meters inserted. The meters are then checked for location and orientation, and concrete is carefully backfilled and compacted around them.

    e. *Monitoring Procedures.*—Readings are taken on the strain meters at predetermined intervals prescribed for each dam. The Wheatstone bridge reading device is connected to the terminals for each strain meter at the terminal board. A standard resistor unit should be used before and after strain observations to confirm proper operation of the test set. Readings are taken and then recorded on a field data form (fig. 4-20).

    f. *Maintenance.*—The current which activates the readout device is supplied by flashlight batteries contained in the test set box. These batteries should be replaced every 6 months or sooner, or whenever the galvanometer needle deflections become weak or sluggish. Holding the galvanometer switch button down continuously while balancing the bridge should be avoided because this causes an excessive drain on the batteries as well as raising the temperature of the embedded instrument resistance coils.

    At least annually, or whenever there is reason to suspect any malfunctioning of the test set, the set should be partially dismantled, accumulated dust or dirt removed from the internal parts, and contact points should be cleaned. Any extensive repairs or recalibration should only be done by the manufacturer.

    g. *Data Processing and Review.*—The field data forms should be forwarded to the Denver Office for processing and review as indicated in section 1.25. Data plots are then prepared as illustrated on figure 4-21.

4.8 Joint Meters.—a. *General.*—Joint meters are used to measure relative movements across a joint, such as the joint between two monoliths. Joint meters operate in a manner similar to strain meters (sec. 4.7). Temperature changes and their effects on the strain or movement are also monitored.

    b. *Advantages and Limitations.*—Joint meters, usually of the Carlson type, have been successfully used for many years. These meters require care in installation and in obtaining observations. Joints that may have a relative movement greater than 0.2 inch (5 mm) should not be measured by joint meters because they have a relatively smaller range of observable movement. The meters are encased in brass to minimize corrosion problems.

    c. *Description of Devices.*—The joint meter (fig. 4-22) resembles the strain meter with the exception that a hub is provided on one end of the joint meter to fit a tapped socket. The socket end is embedded in the concrete of a leading monolith to form the anchorage for the meter on one side of the joint to be measured, while the flange on the other end of the meter is held by the concrete in the monolith on the opposite side of the joint.

    Inside the brass case, a steel framework supports ceramic pulleys and long single loops of steel wire that are held in tension by small coil springs. The greater part of the displacement that the meter experiences is taken up by the springs that provide for movements in the standard joint meter of about 0.2 inch (5 mm). Other models are available with ranges that are either larger or smaller than 0.2 inch.

    The joint meter is constructed so that a 0.01-percent change in resistance ratio of the coils occurs for each 0.0005-inch (0.013-mm) increase in length of the meter. The series resistance of the two coils is about 55 ohms at 70 °F (21.1 °C), and the resistance changes by about 1 ohm for each 11 °F (6.1 °C) change in temperature. The correction for thermal expansion of the meter frame is negligible compared to the magnitude of the movements being measured and is usually not considered in the computations for joint opening.

    The cables, terminal board, and readout devices used with the joint meter are the same as those discussed in section 4.7 for the strain meter.

    d. *Installation Procedures.*—Whenever possible, the joint meter installation (fig. 4-22) should be so arranged that the meter unit and cable lead are placed in a following block. While other arrangements are possible, they require more care in placement and frequently involve field splicing and/or special protection facilities for the meter and cable. The most convenient location for joint meters is 6 to 10 inches (152 to 254 mm) below the top of a concrete lift.

UNITED STATES
DEPARTMENT OF THE INTERIOR
BUREAU OF RECLAMATION
MISSOURI RIVER BASIN PROJECT
YELLOWTAIL DAM
METER NOTES

OBSERVER _L.C._ RECORDER _L.C._ TIME 8:40 AM to 10:00 AM DATE 3-16-82

METER	TOTAL RESIST- ANCE	RESIST- ANCE RATIO	REVERSE RATIO	METER	TOTAL RESIST- ANCE	RESIST- ANCE RATIO	REVERSE RATIO	REMARKS:
S751	70.26	1.0067	.9933	S781	70.38	.9992	1.0008	
S751X	70.02	1.0072	.9928	S781X	70.02	1.0132	.9869	
S752	69.93	1.0136	.9865	S782	69.98	1.0122	.9879	
S752Y	70.00	1.0226	.9778	S782Y	70.28	1.0055	.9945	
S753	70.01	1.0065	.9935	S783	70.39	1.0078	.9922	
S753Z	70.10	1.0114	.9887	S783Z	70.84	1.0195	.9806	
S754	70.17	1.0106	.9895	S784	70.69	1.0067	.9933	
S755	69.91	1.0050	.9950	S785	70.96	1.0174	.9828	
S756	70.28	1.0072	.9928	S786	70.54	.9989	1.0011	
S757	70.24	.9959	1.0041	S787	70.23	1.0142	.9859	
S758	70.15	1.0029	.9971	S788	70.55	.9975	1.0025	
S759	70.22	1.0001	.9999	S789	70.80	1.0259	.9747	
S761	71.02	.9961	1.0039	S791	69.50	1.0135	.9866	
S761X	70.88	.9987	1.0013	S791X	70.14	1.0154	.9848	
S762	70.82	1.0035	.9965	S792	69.67	1.0291	.9717	
S762Y	70.77	1.0042	.9958	S792Y	69.59	1.0211	.9793	
S763	70.81	1.0109	.9892	S793	69.73	1.0189	.9814	
S763Z	70.89	1.0073	.9927	S793Z	69.81	1.0126	.9875	
S764	71.02	.9967	1.0033	S794	69.67	1.0224	.9780	
S765	70.80	1.0001	.9999	S795	69.53	1.0062	.9938	
S766	70.91	1.0270	.9737	S796	69.52	1.0008	.9992	
S767	70.97	1.0068	.9932	S797	69.80	1.0018	.9982	
S768	70.96	.9995	1.0005	S798	69.71	1.0035	.9965	
S769	71.14	1.0106	.9895	S799	69.93	1.0295	.9714	
S771	70.58	1.0013	.9987	NS21	78.59	1.0184	.9819	
S771X	70.67	.9985	1.0015	NS22	78.22	1.0148	.9854	
S772	70.64	1.0035	.9965	J61	59.14	.9943	1.0057	
S772Y	70.78	.9989	1.0011	J62	59.73	.9917	1.0083	
S773	70.27	1.0081	.9919	J63	60.34	.9959	1.0041	
S773Z	70.70	1.0048	.9952	J64	60.33	.9903	1.0097	
S774	70.61	1.0041	.9959	J65	59.60	.9979	1.0021	
S775	70.62	1.0055	.9945	J66	58.83	.9983	1.0017	
S776	70.56	1.0136	.9865					
S777	70.79	1.0068	.9932					
S778	70.58	1.0062	.9938					
S779	70.47	.9945	1.0055					

STD. COIL	BEGIN	END	TEST SET SERIAL NO.
RES. RATIO	1.0302	1.0302	23495
REV. RATIO	.9707	.9707	

Figure 4-20.—Field data form for a strain meter. 459-D-906.

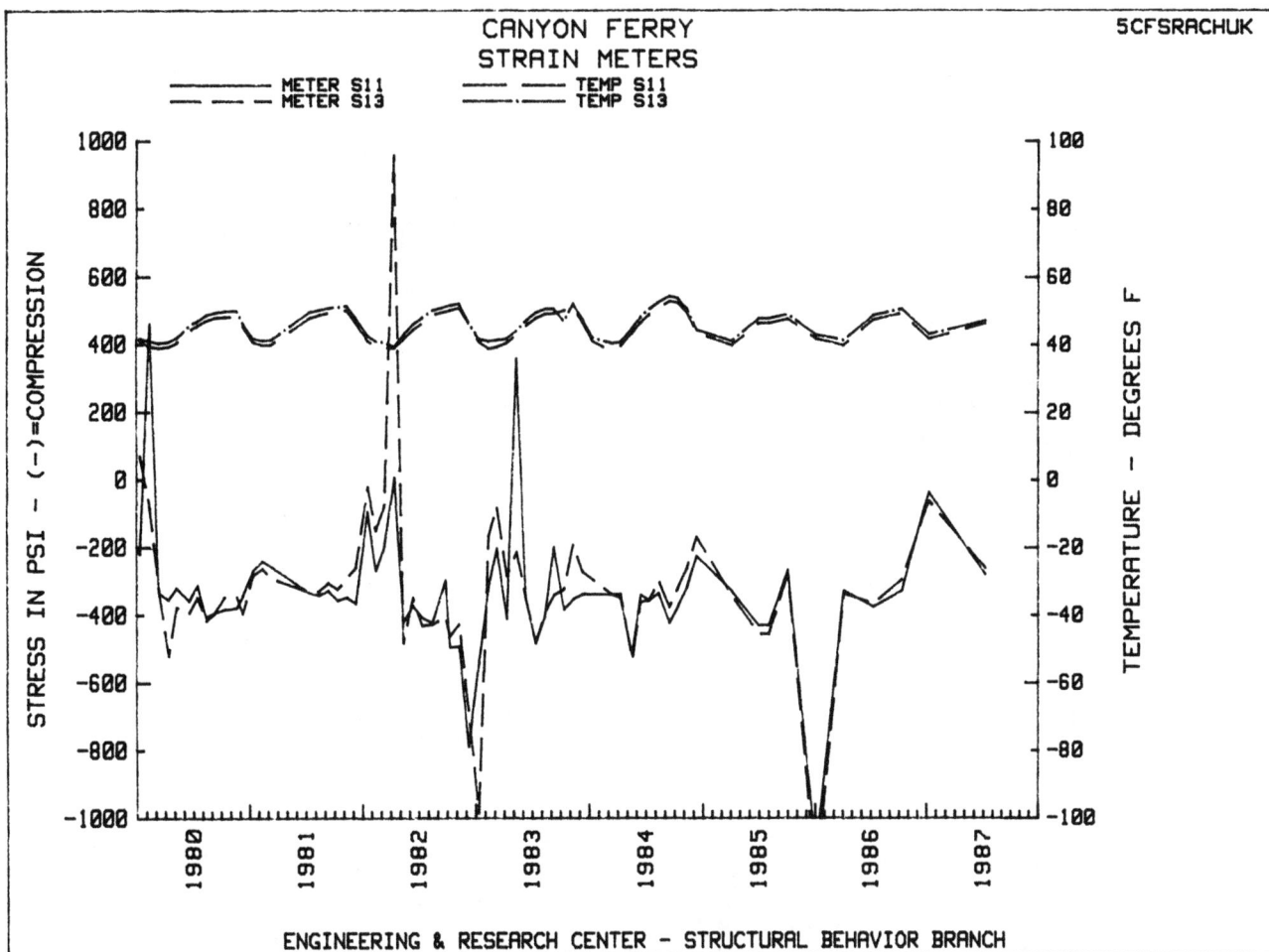

Figure 4-21.—Data plots on strain meters.

A typical installation procedure for a joint meter would be:

1. Socket plug is nailed to interior surface of form or block-out at location desired; slotted side is then placed against form.
2. Meter socket is then screwed into the plug, completing preliminary portion of meter installation. Concrete is then placed inside the form.
3. After concrete in following block is brought up to the elevation of the joint meter location, a small portion of the fresh concrete is removed to a depth necessary to expose plug in end of socket embedded in leading block.
4. Remove plug and screw in joint meter unit tightly. The meter is placed into a position that will provide sufficient available operational range to cover magnitude and direction of expected joint movement. Normally, a completely closed or mid-range position will be satisfactory.
5. The initial meter position is determined by making ratio readings with a portable readout device.
6. Concrete is then carefully backfilled and compacted around the device.

   e. *Monitoring Procedures.*—Procedures for monitoring joint meters are the same as those described in section 4.7 for strain meters.
   f. *Maintenance.*—Required maintenance on joint meter installations is accomplished only on the readout device as described in section 4.7.
   g. *Data Processing and Review.*—The field data are entered on a data form (fig. 4-23), and then forwarded to the Denver Office as described in section 1.25. Data plots are then prepared as illustrated on figure 4-24.

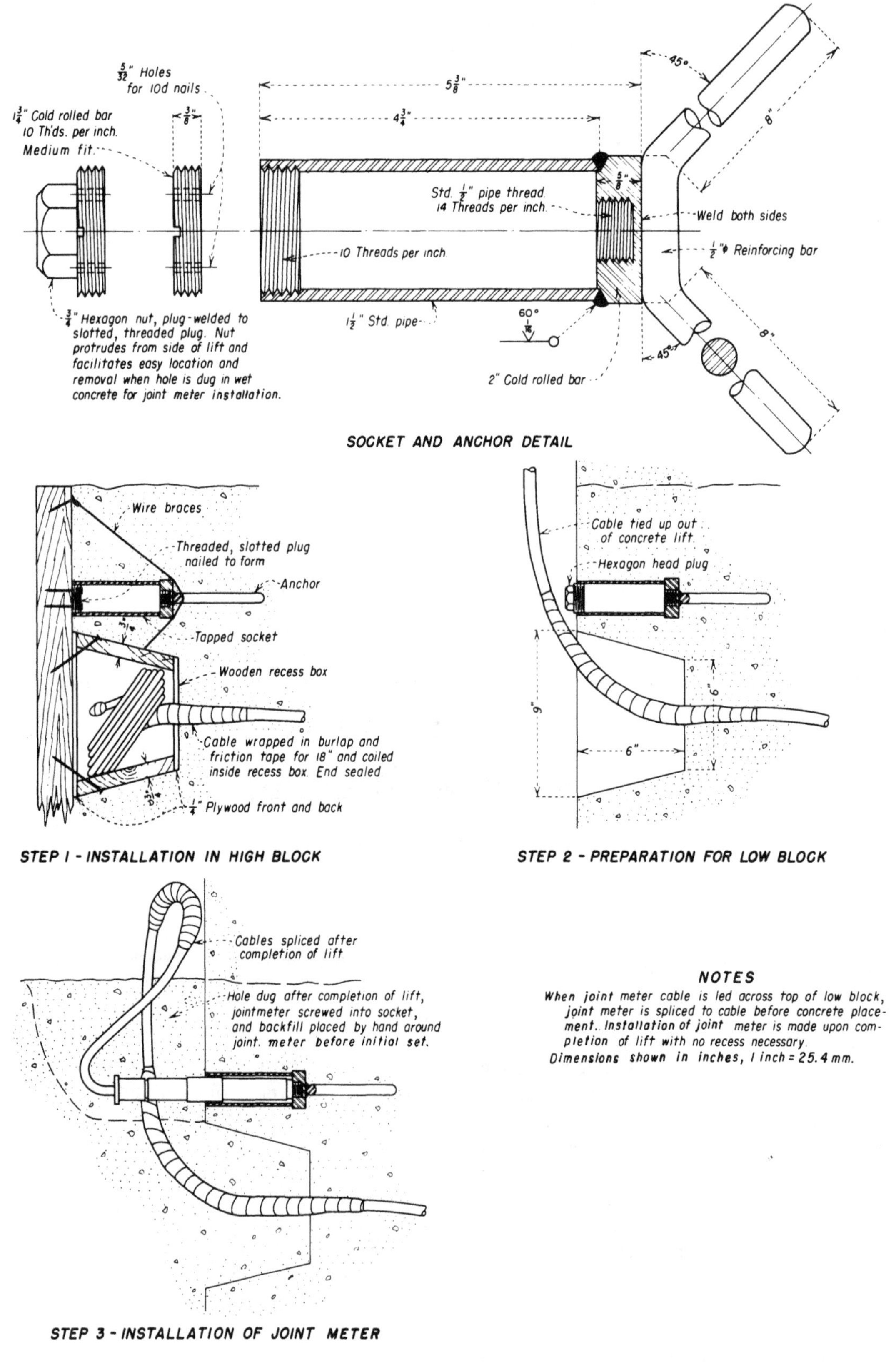

SOCKET AND ANCHOR DETAIL

STEP I - INSTALLATION IN HIGH BLOCK

STEP 2 - PREPARATION FOR LOW BLOCK

STEP 3 - INSTALLATION OF JOINT METER

NOTES

When joint meter cable is led across top of low block, joint meter is spliced to cable before concrete placement. Installation of joint meter is made upon completion of lift with no recess necessary.

Dimensions shown in inches, 1 inch = 25.4 mm.

Figure 4-22.—Typical joint meter installation. 40-D-5094.

78

4.9 Foundation Deformation Meters.—a. *General.*—Foundation deformation meters are installed to measure movement of the foundation rock relative to the mass of the dam. These meters are usually installed in drill holes as shown on figure 4-25.

   b. *Advantages and Limitations.*—Foundation deformation meters have been successfully used on several Bureau structures such as Yellowtail, Pueblo, and Morrow Point Dams. Obviously, care in installation and taking readings is of extreme importance, as is proper placement of the devices to ensure securing the necessary data required.

   c. *Description of Devices.*—The foundation deformation meter is similar to the joint meter previously described except that the socket end has a length of steel pipe attached (fig. 4-25). This pipe, which is inserted into a drill hole, is of sufficient length to extend into the foundation stratum where movement relative to the dam is to be measured. All other portions of the device are the same as described in section 4.8 c. A second device for measuring foundation deformation, which uses an invar tape and a micrometer instead of a joint meter, is shown on figure 4-26.

   d. *Installation Procedures.*—After a drill hole is made and cleaned to the depth desired, the pipe is inserted into the hole and the bottom 5 feet (1.5 m) is grouted solidly into place. The portion of the pipe not grouted is greased to prevent bonding during foundation grouting. Just prior to concrete placement, the deformation joint meter is carefully installed and embedded in the first lift of the dam.

   e. *Monitoring Procedures.*—Readings on deformation meters are made in the same manner as for joint meters; the readings are recorded on a field data form as shown on figure 4-27. A typical layout for these meters is shown on figure 4-28.

   f. *Maintenance.*—Maintenance on the readout device is identical to that described in section 4.7 f for strain meters.

VIBRATING-WIRE JOINT METERS

DAM Brantley

OBSERVER(s) John Suggs

RESERVOIR ELEVATION

DATE   03/13/87

TAILWATER ELEVATION

SHEET   5   OF   7

JOINT METER NUMBER	SERIAL NUMBER	GAUGE FACTOR	INITIAL READING (Freq)	CURRENT READING (Freq)	DEFORMATION (INCHES)	TEMPERATURE CORRECTION			DEFORMATION CORRECTED FOR TEMP. (IN.)
						FACTOR	INITIAL TEMP.	TEMPERATURE	
		G	F_o	F_1	*D	C	T_o (°F)	T_1 (°F)	**D_T
JTUR01	JTUR01	0.00042	1766	2604	0.35196	0.00025	76.6	59.8	0.34776
JTDR01	JTDR01	0.00034	1839	2857	0.34612	0.00025	73.0	62.1	0.34340
JTUR02	JTUR02	0.00034	1711	2642	0.31654	0.00025	69.6	58.6	0.31379
JTDR02	JTDR02	0.00034	1850	No Reading		0.00025	71.6	No Reading	
JTUR03	JTUR03	0.00035	1728	2562	0.29190	0.00025	72.8	56.1	0.28773
JTDR03	JTDR03	0.00036	1555	2569	0.36504	0.00025	71.9	59.2	0.36187
JTUR04	JTUR04	0.00036	1594	2570	0.35136	0.00025	68.4	58.5	0.34889
JTDR04	JTDR04	0.00034	1573	1952	0.12886	0.00025	65.4	62.3	0.12809

*$D=G(F_1-F_o)$
**$D_T=G(F_1-F_o)+C(T_1-T_o)$

Figure 4-23.—Field data form for joint meters.

GRAND COULEE FOREBAY DAM
JOINT METERS

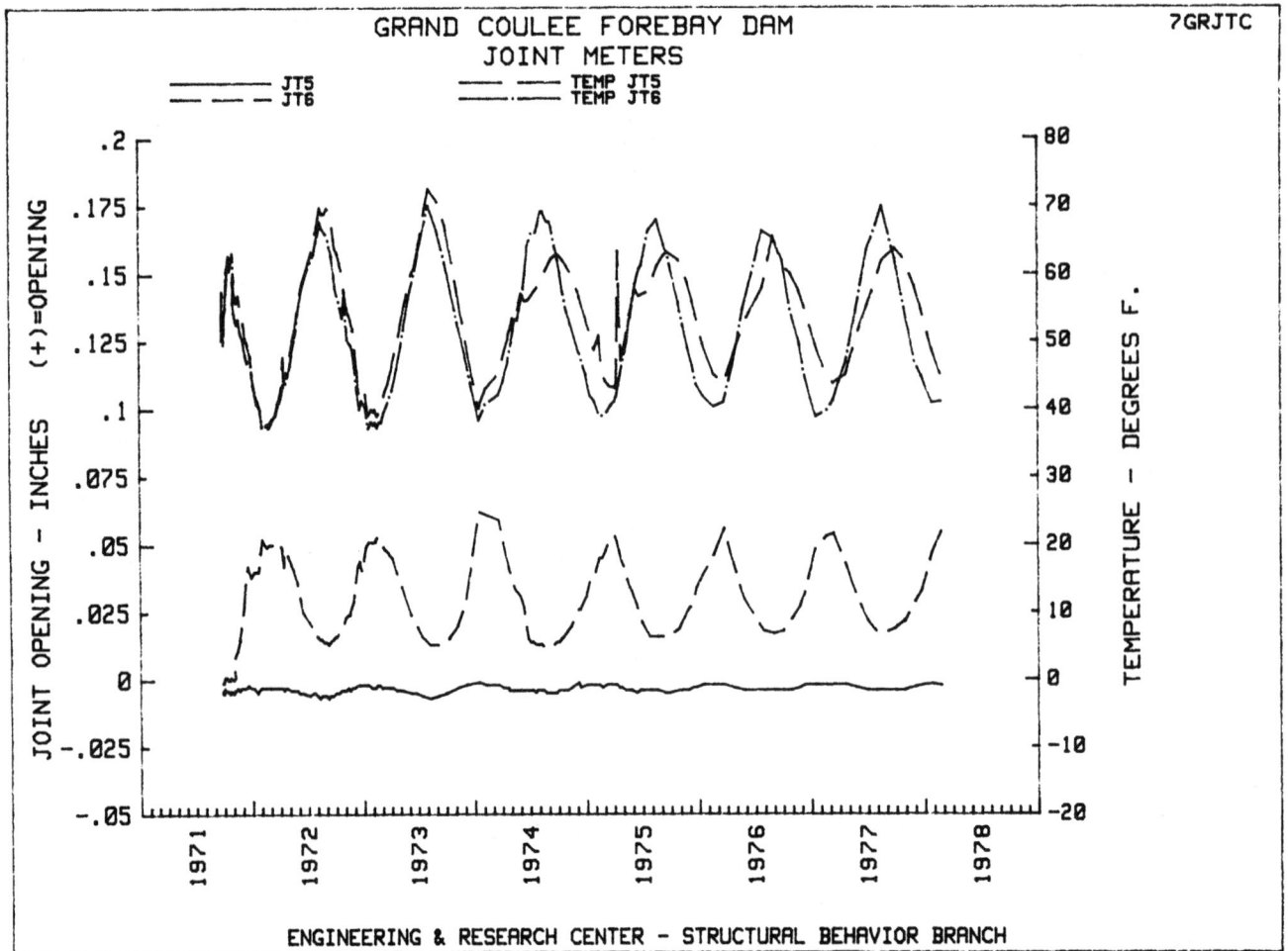

Figure 4-24.—Data plots on joint meters.

g. *Data Processing and Review.*—Field data forms are forwarded to the Denver Office for processing and review as described in section 1.25. Data plots are then prepared as shown on figure 4-29.

4.10 "No Stress" Strain Meters.—a. *General.*—A more complete analysis of stress within a dam requires data regarding the volume changes that take place in the concrete in the absence of loading. These volume changes may be the result of volumetric effects of temperature and moisture changes and chemical action within a large structure. "No stress" strain meters, usually used to obtain these data, have been provided in conjunction with other groups of strain meters in some dams.

b. *Advantages and Limitations.*—As previously stated, no-stress strain meters provide the necessary data that allows the separation of changes due to stress and changes due to nonload related movements. These devices must be installed with great care, and the analysis performed by trained competent personnel.

c. *Description of Devices.*—The devices used are the same as for the strain meters previously described in section 4.7 c; however, the installation procedures are different.

d. *Installation Procedures.*—The objective of the installation of a no-stress strain meter is to embed the meter in mass concrete that is isolated from any deformation due to loading, but is responsive to prevailing temperature, moisture, and growth changes in the concrete. One method that has been used successfully is to provide a cavity in the lift near a strain meter group by embedding a 3- or 4-foot (0.9- or 1.2-m) length of 15-inch (381-mm) concrete pipe so that the flange of the pipe extends about 3 inches (76 mm) above the lift surface. A concrete pedestal is constructed around the bottom of the pipe to support and hold the pipe in place during placement of the concrete lift. A precast concrete cover is also provided for the pipe. The strain meter is suspended in the center of a 12- by 24-inch (305- by 610-mm) cylinder specimen mold of a type that will permit easy stripping. The mold is then filled with concrete taken from a typical lift adjacent

80

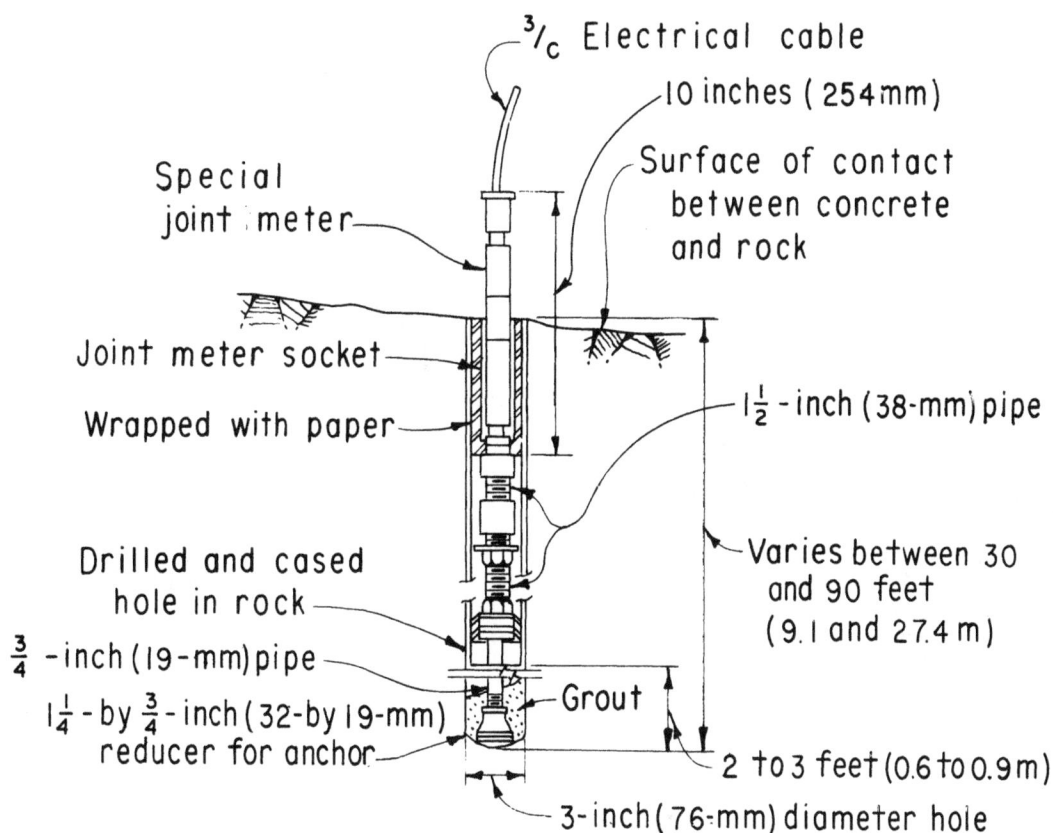

Figure 4-25.—Foundation deformation meter installation. 288-D-3095.

to the meter group. The day after the cylinders are cast, the molds are removed, cavity cleaned, concrete cylinder placed in cavity, and the precast cover placed and sealed.

Another method that has been used successfully by the Bureau is to embed two strain meters, one vertically and one horizontally near the top of a concrete lift in an isolated cone of concrete. Then, a 3-foot (0.9-m) diameter, ⅜-inch (9.5-mm) thick steel plate is placed over the lift surface above the meter location. The plate is held 2 to 3 inches (51 to 76 mm) above the concrete cone surface by a circumferential rim or lip that isolates the meters from the effects of vertical load. The length change indicated by the vertical meter includes the deformations resulting from the horizontal strains modified by Poisson's ratio; the horizontal strain meter aids in the evaluation of the effect of Poisson's ratio. Figure 4-30 illustrates such an installation using a truncated cone of concrete.

A third method, developed by J. L. Serafim, involved the use of a double-walled, double-bottomed, copper container that serves to protect the concrete containing the meter from strains due to load, yet provides continuity of the concrete so that other volumetric changes may take place without restraint.

    e. *Monitoring Procedures.*—The monitoring procedure for no-stress strain meters is the same as that described in section 4.7 e for common strain meters.

    f. *Maintenance.*—The only maintenance possible is that performed by cleaning the terminal board and checking the readout unit as described in section 4.7 f.

    g. *Data Processing and Review.*—The field data are transmitted to the Denver Office for analysis and review as discussed in section 1.25. Data plots are prepared as shown on figure 4-31.

4.11 Optical Plummets.—**a. *General.***—Optical plummets represent a variation of the previously discussed plumbline type of installation wherein horizontal movements of a dam are determined with reference to the dam foundation.

    b. *Advantages and Limitations.* An optical plummet system has the advantage over a plumbline system in that none of the plumbline equipment is required; therefore, there is no concern over deterioration of that equipment. However, a very precise setup is required for reading an optical plummet. The system used is

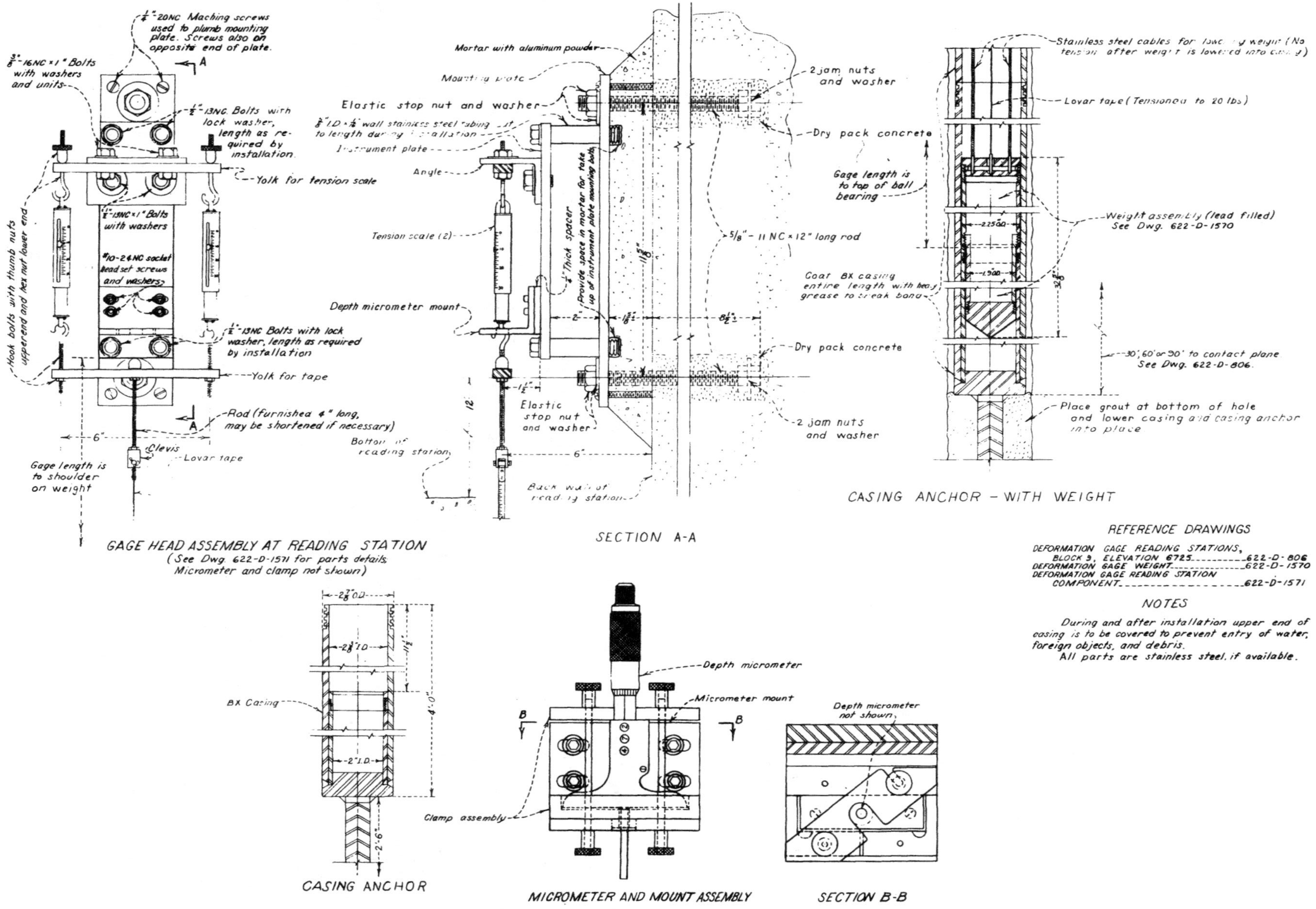

Figure 4-26.—Details of a foundation deformation meter. 622-D-938.

82

CRYSTAL DAM
FOUNDATION DEFORMATION METER NOTES

Observer *POTTER* Recorder *TAYLOR*   Time *11:50* to *12:22* Date *12-12-84*

RES. EL. 6746.58

METER	TOTAL RESISTANCE	RESISTANCE RATIO	REVERSE RATIO	REMARKS
D 1	60.65	103.42		
D 2	58.72	100.86		
D 3	59.66	101.52		
D 4	—	—		*BAD READING- METER JUMPS*
D 5	60.07	99.95		
D 6	60.01	101.57		
D 7	—	—		*BAD READING- METER JUMPS*
D 8	59.75	102.24		
D 9	58.75	100.76		
D 10	58.68	99.19		
D 11	59.06	98.70		
D 12	59.67	97.67		
D 13	59.10	101.53		
D 14	59.46	100.50		
D 15	58.62	100.54		
D 16	58.46	100.97		

STANDARD COIL	BEGIN	END	TEST SET SERIAL NO. 7808017
RESISTANCE RATIO	67.00	67.02	
REVERSE RATIO	103.03	103.03	

Figure 4-27.—Field data form for foundation deformation meters. 622-D-3056.

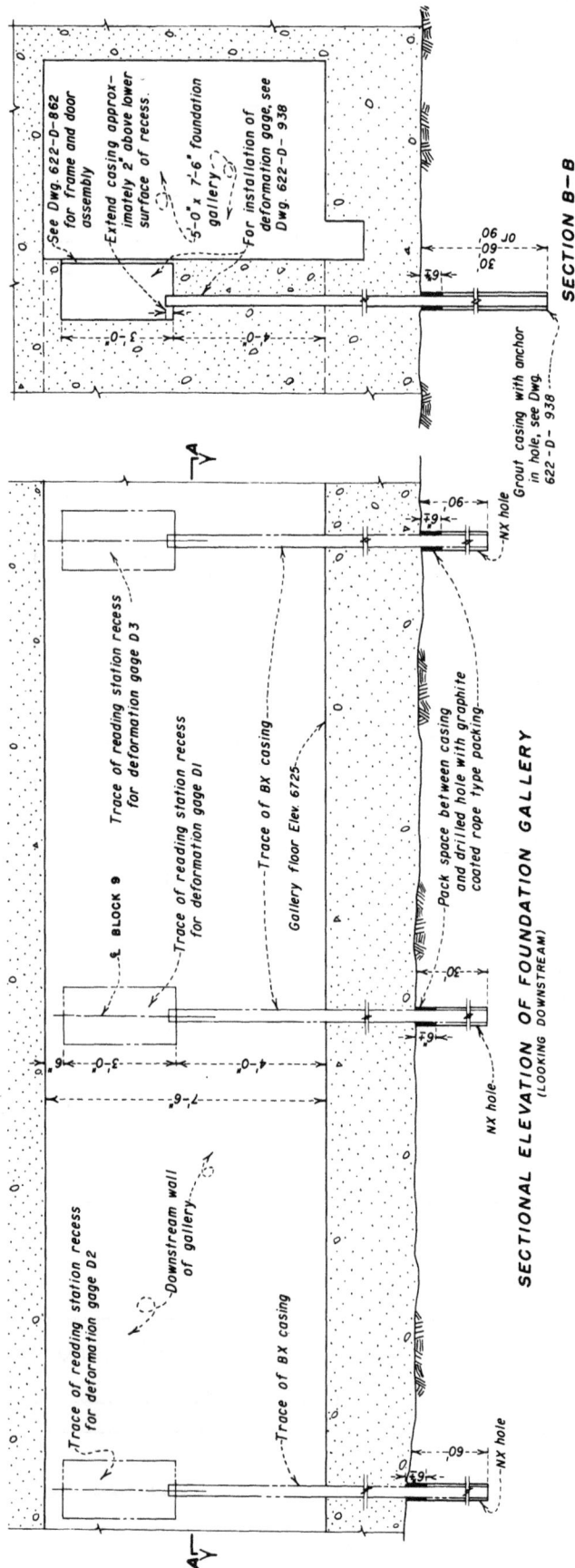

SECTIONAL ELEVATION OF FOUNDATION GALLERY
(LOOKING DOWNSTREAM)

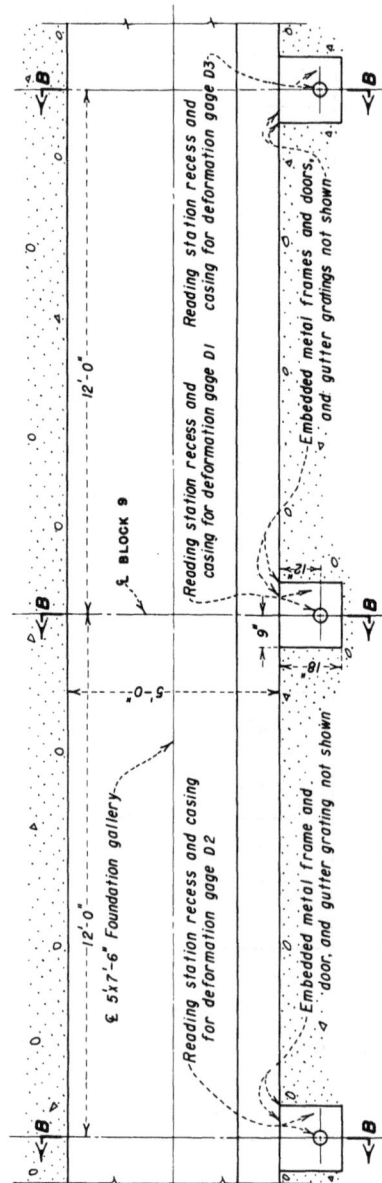

SECTION B-B

SECTIONAL PLAN A-A

NOTES

Wrap BX casing for each deformation meter between foundation line and recess with burlap and coat with heavy grease to break bond with the concrete.
For installation of typical deformation gage—see Detail M. Drawing 622-D-804.

Figure 4-28.—A typical reading station layout for foundation deformation meters. 622-D-806.

84

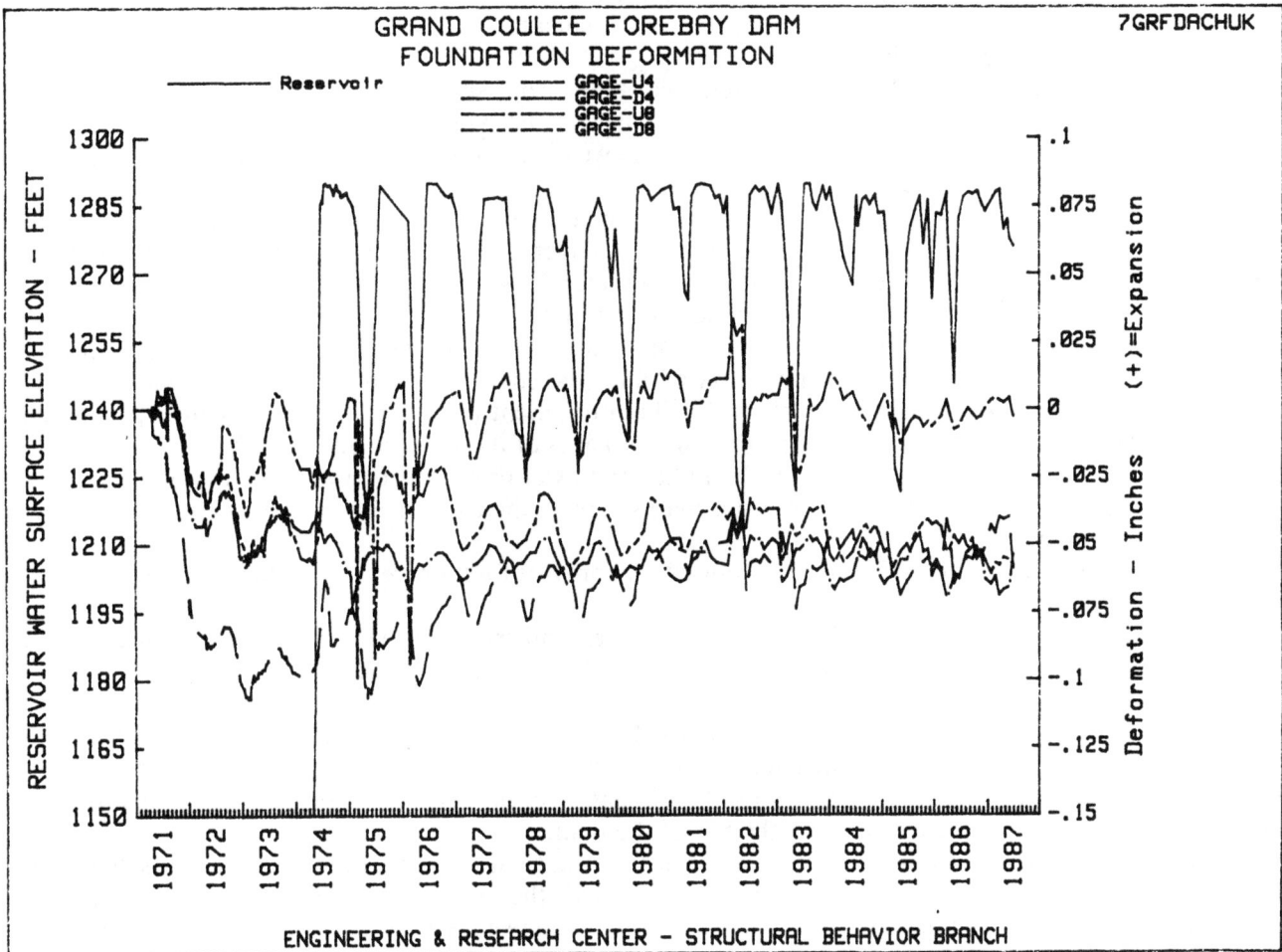

GRAND COULEE FOREBAY DAM
FOUNDATION DEFORMATION
7GRFDACHUK

Figure 4-29.—Data plots on foundation deformation meters.

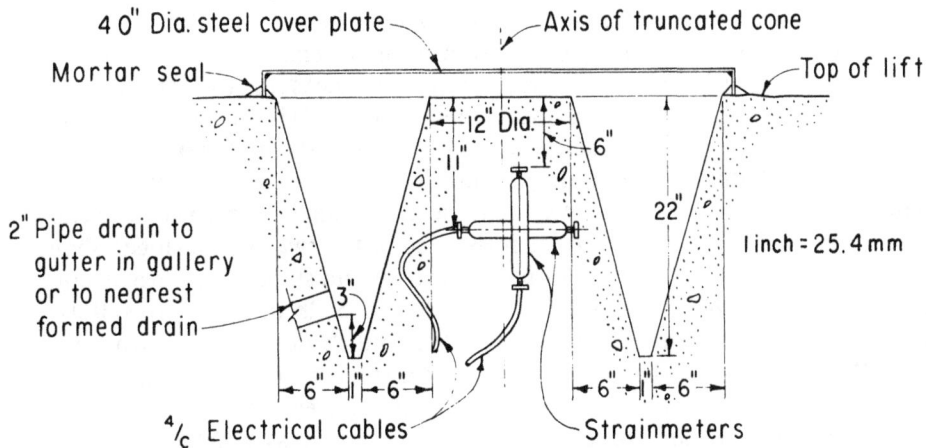

Figure 4-30.—Sectional elevation of a no-stress strain meter installation. 288-D-3091.

85

accurate to ±0.005 foot (±0.0015 m), which means that the error introduced in this type of instrumentation could be greater than the magnitude of the desired measurement. A conscientious, well-trained observer is necessary to obtain accurate readings.

c. *Description of Devices.*—The optical plummet system consists of a vertical well through the dam (generally into the foundation), a reference target installed at the base of the well, and a tripod-mounted plummet with a reference grid at the top of the hole. The well is illuminated so that the reference target can be visually referenced.

d. *Installations Procedures.*—A vertical well is constructed through the dam in the same manner as that described in section 4.3 d for plumblines, and a reference target is installed in the center of the base of the well, see figure 4-32. The top of the well is fitted with a meter flange marked so that a transparent reference grid can be positioned in the exact center of the top of the well.

e. *Monitoring Procedures.*—Optical plummet monitoring is accomplished by positioning and leveling a tripod-mounted plummet over the well so that a line of plumb is established between the instrument crosshairs and the center of the target at the base of the well. The transparent plate with its grid system is positioned at the floor elevation at the top of the well so that the zero axes of its grid coincide with the reference marks on the flange of the well. The differences in the radial and axial directions between the line of plumb and the origin of the lines of the grid on the reference marks furnish the components of deformation at the time of each measurement.

f. *Maintenance.*—The only maintenance required is to provide for the care of the optical plummet reading device.

g. *Data Processing and Review.*—The field data are transmitted to the Denver Office for analysis and review as discussed in section 1.25. Data plots are prepared as shown on figure 4-33.

4.12 Stress Meters.—a. *General.*—Stress meters are used for such special applications as determining vertical stresses at the bases of sections for comparison and as checks of stresses computed from strain meter results. In arch dams, stress meters may also be used for determination of horizontal stresses normal to the direction of thrust in the thinner arch elements near the top of the dam.

b. *Advantages and Limitations.*—A stress meter is designed to measure only compressive stress; however, because of the manner in which stress meters are calibrated, they may also exhibit some tension. This is accomplished by establishing the zero stress value after the concrete has set (usually 24 hours). In this manner, the stress meter may subsequently be unloaded, and will indicate an apparent tension. Values from stress meters are usually analyzed using a computer program because of the relative complexity of the data reduction.

c. *Description of Devices.*—The details of the construction of a stress meter are shown on figure 4-34. The meter consists of a mercury-filled diaphragm that is shaped like a plate, with a chamber that contains the measuring unit protruding from one side. The center of the plate under the measuring unit chamber is somewhat flexible because of a cavity in the center of the plate. The mercury that is in contact with the plate causes the plate to deflect elastically in direct proportion to the intensity of pressure on the diaphragm. The measuring unit consists of a small elastic wire strain meter that measures the intensity of stress as a change in the resistance ratio of the two coils of the strain meter, and measures the temperature in terms of the series resistance of the two coils, much in the same manner as the strain meter.

The range of the standard stress meter is from 0 to 800 lb/in^2 (0 to 5.5 MPa) in compression, with a sensitivity of about 5 lb/in^2 (0.24 MPa) for every 0.01 percent change in resistance ratio. Stress meters are also available in other pressure ranges. The series resistance of the two coils changes about 1 ohm for every 8°F (4.5°C) change in temperature. A temperature correction must be made when computing the stress indicated by the stress meter. The correction is from 0.5 to 1.5 (lb/in^2)/°F (0 to 0.01 MPa/°C) change in temperature. The meter readings are taken by a similar readout device used for strain meters.

d. *Installation Procedures.*—Satisfactory operation of a stress meter is dependent almost entirely upon obtaining complete contact between the meter plate and the adjacent concrete. The installation procedure used must avoid the formation of air pockets and eliminate, as far as practicable, the collection of water beneath the meter that may result from bleeding.

Stress meters may be placed in the horizontal position (with stem vertical) or in a diagonal position (with stem sloping) as outlined in the following procedures:

1. Prior to completing placement of concrete in the lift in which the meter is to be embedded, the stress meter and about 4 feet (1.2 m) of cable are placed in a wooden box only slightly larger than the meter.

86

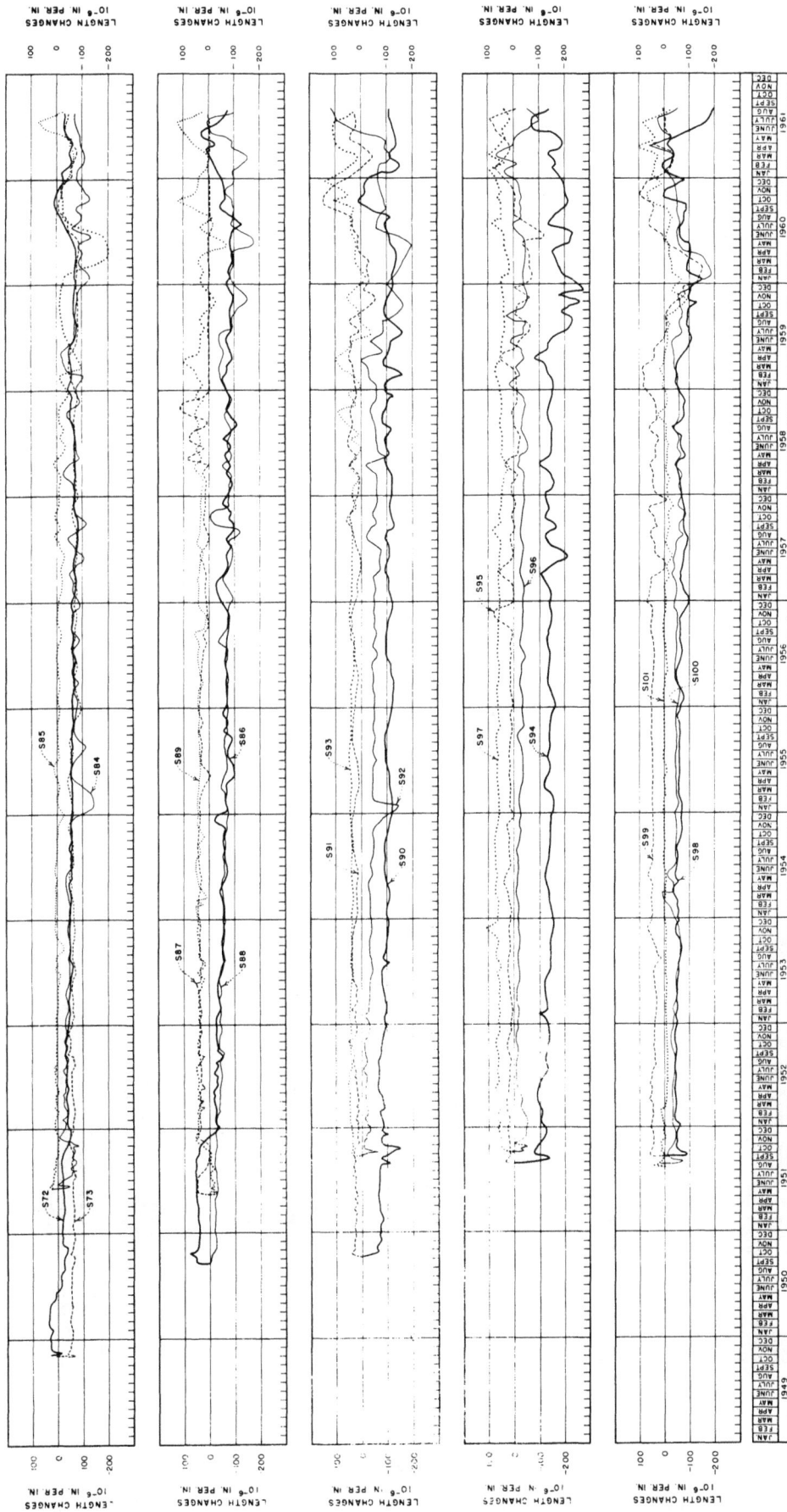

Figure 4-31.—Data plots on a no-stress strain meter.

SIGN CONVENTION

POSITIVE SIGN INDICATES INCREASE IN LENGTH

87

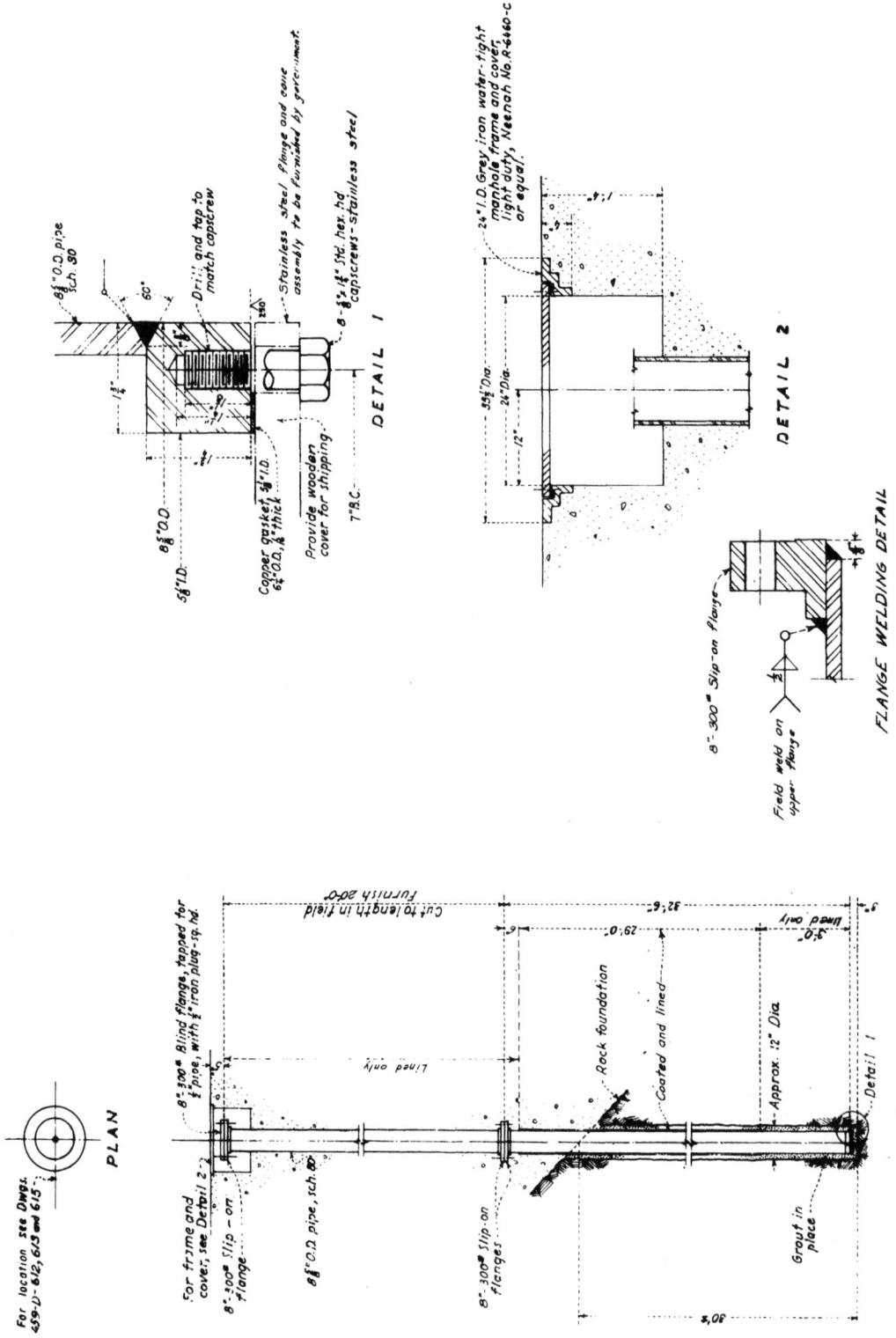

Figure 4-32.—Typical installation details for an optical plummet well. 459-D-711.

88

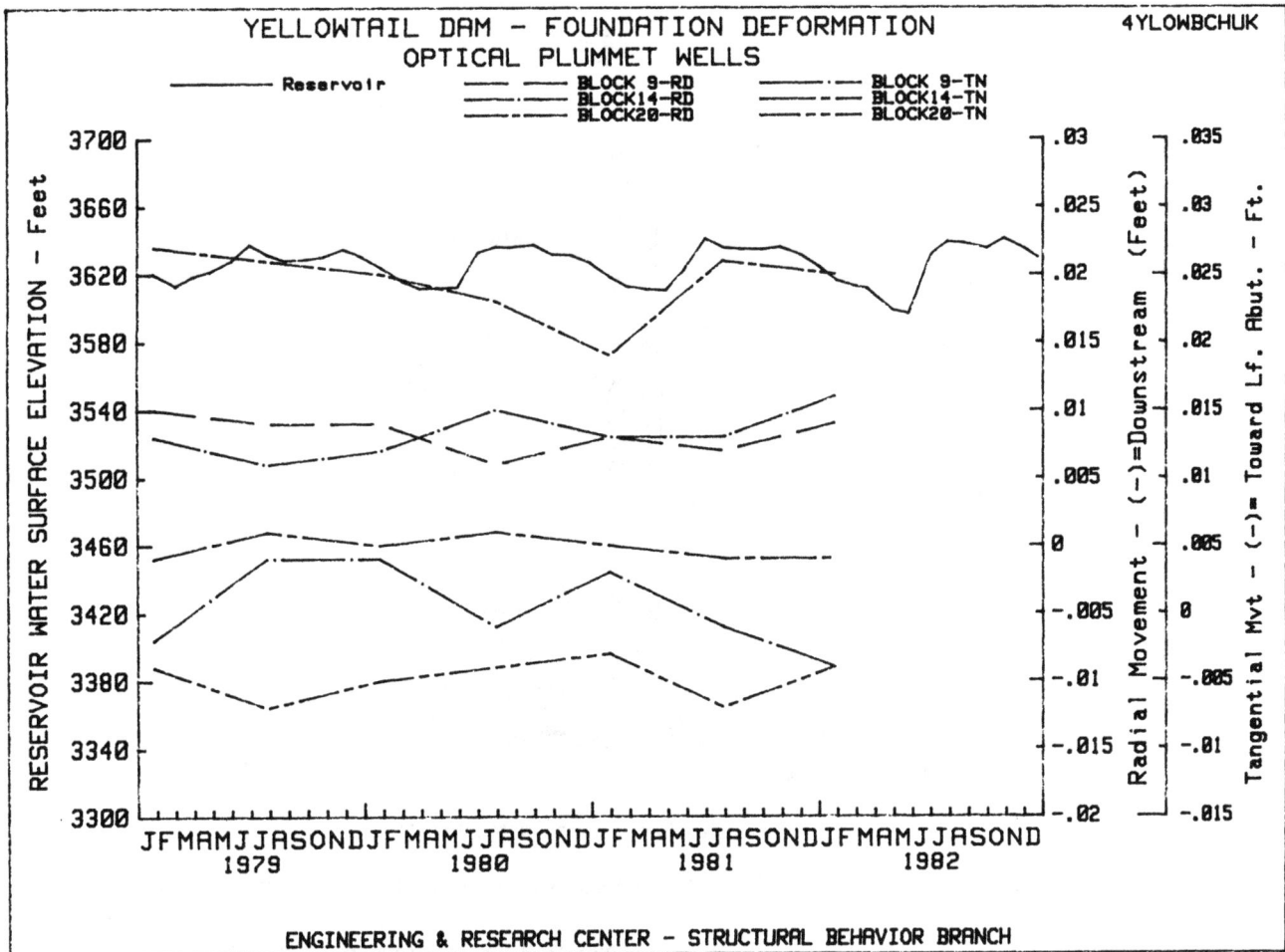

Figure 4-33.—Data plots on optical plummet wells.

2. When concrete has reached an elevation about 12 inches (305 mm) below the top of the proposed lift, the box containing the meter is placed at the desired location and the lift placement completed. The meter cable leads are placed and buried during this period in the usual manner except for the extra cable in the box.

3. Upon completion of concrete lift, the box containing the meter is removed from the concrete and a conical cavity is left in the lift surface. The side slopes and bottom of this cavity should be sloped or leveled as required into relatively smooth plane surfaces. Excessive troweling should be avoided. Figure 4-35 illustrates this installation process.

4. The following day, after the concrete has hardened, the hole is cleaned to remove all loose material and water. Projecting aggregate corners are chipped away and brushed to ensure a clean surface.

5. About 1 hour before placing the meter, a grout is prepared consisting of about 3 ounces (80 grams) of cement and 4 ounces (120 grams) of sand finer than the No. 30 (600-μm) sieve. Only enough water to produce a plastic consistency should be used. The grout is then placed in the conical cavity to receive the meter.

6. Meter is pressed into the grout with a reciprocal rotary motion that will cause the grout to squeeze out from around the base of the meter. The grout bed should not exceed ⅛ inch (3 mm) in thickness. Weights are then applied to hold meter in place.

7. After grout has set for 2 to 3 hours, fresh grout is placed and carefully hand compacted around the meter as the weights are removed. When hole is completely backfilled, it should be protected from traffic until grout has thoroughly set.

Stress meters may also be placed in a vertical position (with stem horizontal) within the fresh concrete near the top of a lift (fig. 4-17):

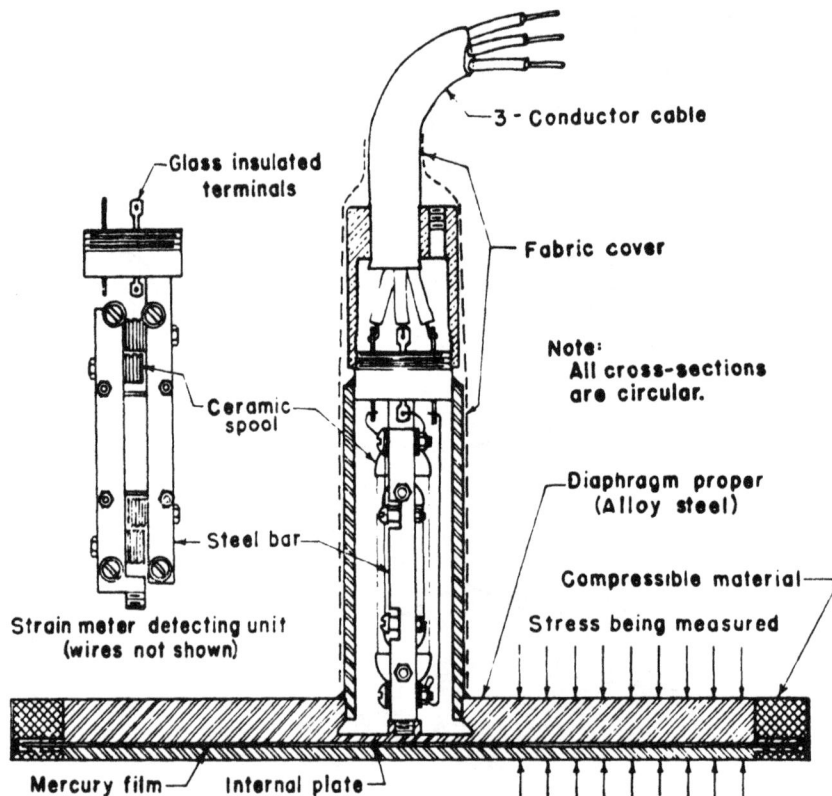

Figure 4-34.—Typical details of a stress meter. From [10].

1. After concrete placement in the lift has been completed, a hole about 12 inches (305 mm) deep is dug at the meter location.

2. Meter is laid in place and fresh concrete, with cobbles removed, is placed in thin layers around the instrument and thoroughly, but carefully, tamped in place.

3. Alignment and position of the meter should be checked to ensure proper position and orientation as backfilling progresses.

4. Area should then be protected until concrete has set.

    e. *Monitoring Procedures.*—The monitoring procedure for stress meters is the same as that discussed in section 4.7 e for strain meters.

    f. *Maintenance.*—The only maintenance possible in the system is occasional cleaning of the terminal board and checking the readout unit as described in section 4.7 f.

    g. *Data Processing and Review.*—The field data forms are transmitted to the Denver Office for analysis and review as described in section 1.25. Data plots are prepared as shown on figure 4-36.

4.13 Other Devices.—a. *General.*—Instruments for measuring stress or strain are sometimes installed in reinforcing steel that surrounds openings in or near a dam such as penstocks, tunnels, spillway openings, and galleries. Instruments may also be installed on rock bolts used to stabilize rock masses and, in some instances, special pore pressure measurement devices may be installed around penstocks. In addition, special temperature sensing devices are frequently installed at various locations within a dam; these devices are discussed in chapter 7.

    b. *Advantages and Limitations.*—In general, the advantages and limitations of each special purpose type of instrumentation are as previously discussed for other types of strain and stress meters.

    c. *Description of Devices.*—The special purpose reinforcing steel and penstock strain meters operate on the principle of changed resistance in a wire as the tension in the wire changes. Figure 4-37 illustrates these devices, and figure 4-38 shows the completed installation of a pore pressure meter installed on the outer surface of a penstock shell.

90

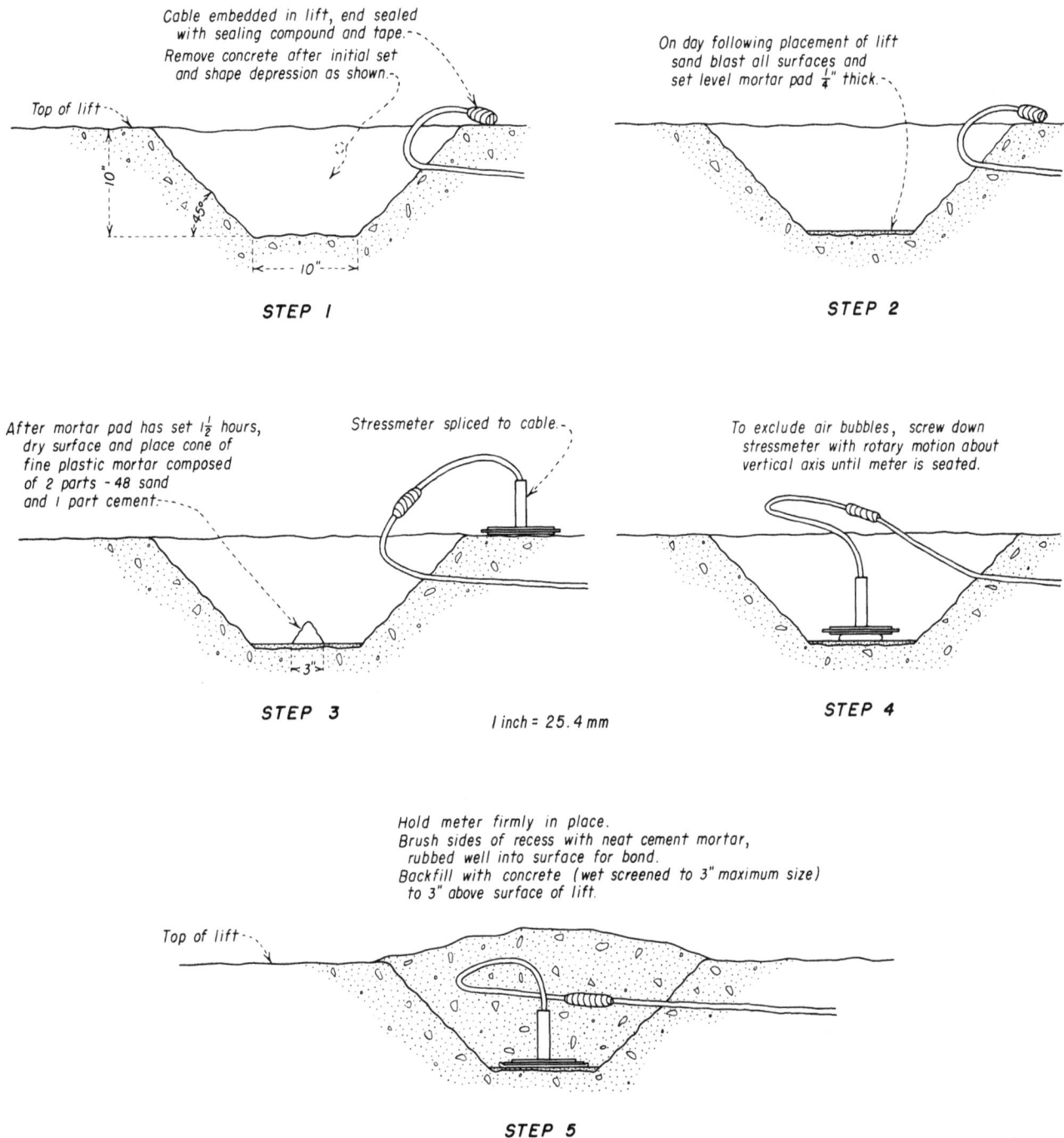

Cable embedded in lift, end sealed with sealing compound and tape.--
Remove concrete after initial set and shape depression as shown.--

Top of lift--

10"

45°

10"

STEP 1

On day following placement of lift sand blast all surfaces and set level mortar pad $\frac{1}{4}$" thick.--

STEP 2

After mortar pad has set $1\frac{1}{2}$ hours, dry surface and place cone of fine plastic mortar composed of 2 parts - 48 sand and 1 part cement.----

3'

STEP 3

Stressmeter spliced to cable.--

1 inch = 25.4 mm

To exclude air bubbles, screw down stressmeter with rotary motion about vertical axis until meter is seated.

STEP 4

Hold meter firmly in place.
Brush sides of recess with neat cement mortar, rubbed well into surface for bond.
Backfill with concrete (wet screened to 3" maximum size) to 3" above surface of lift.

Top of lift--

STEP 5

Figure 4-35.—Installation procedure for vertical stress meters. 40-D-5095.

   d. *Installation Procedures.*—Reinforcing steel strain meters are usually placed on at least one bar of each row of reinforcement at selected locations around the opening to be investigated to measure deformations in the reinforcement. Stress in the reinforcing is computed from the strain measurements.

   Where stress in the steel liners of penstocks is to be investigated, strain meters are attached (fig. 4-37) to the surface of the penstock liner by welded support brackets. Such instruments are usually installed at each of three equally spaced circumferential locations and at two or more elevations on the penstock liner.

   At each location of a penstock strain meter, pore pressure meters are installed at the outer surface of the steel liner to measure possible hydrostatic pressure that may develop between the liner and the surrounding concrete.

91

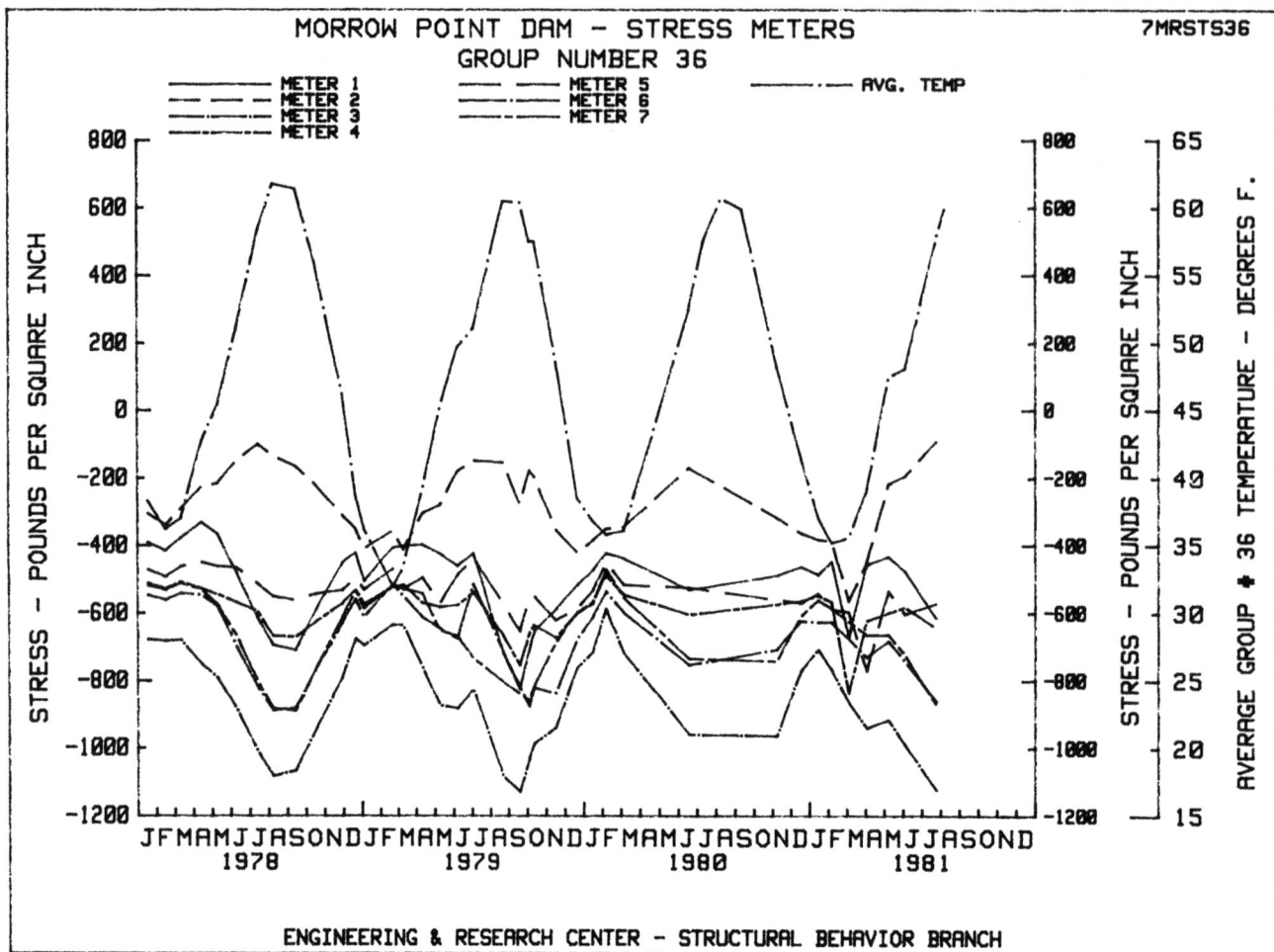

MORROW POINT DAM - STRESS METERS
GROUP NUMBER 36
7MRSTS36

Figure 4-36.—Data plots on stress meters.

e. *Monitoring Procedures.*—The devices are usually read using a similar readout device as discussed in section 4.7 e. Field data forms are prepared as shown on figure 4-39.

f. *Maintenance.*—No maintenance of the devices is possible except for caring for the terminal board and checking the readout device.

g. *Data Processing and Review.*—The field data forms are forwarded to the Denver Office for analysis and review as described in section 1.25. Special data plots are prepared as shown on figures 4-40 and 4-41.

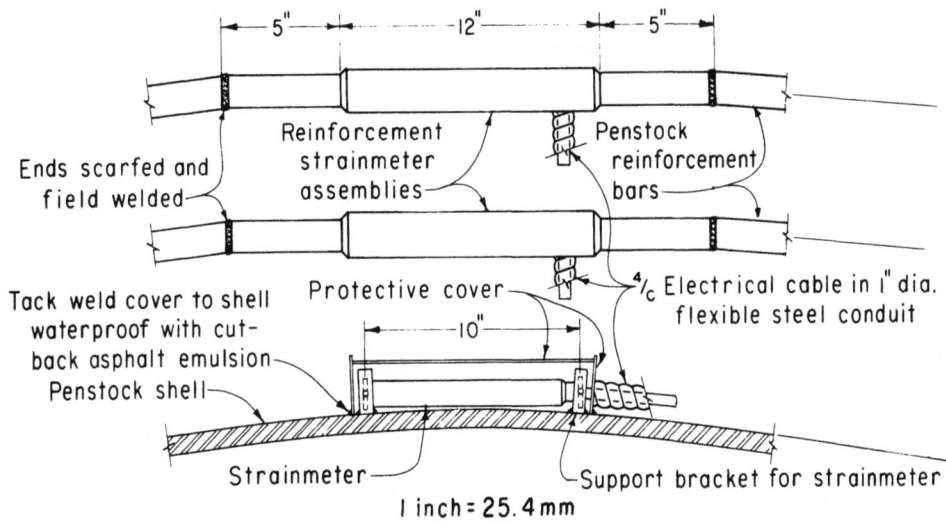

Figure 4-37.—Special purpose reinforcing steel and penstock strain meters. 288–D–3093.

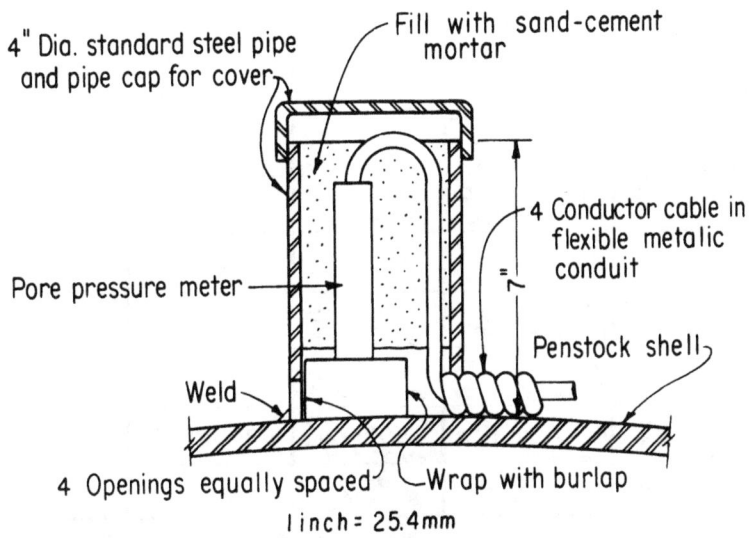

Figure 4-38.—Pore pressure meter installed on outer surface of a penstock shell. 288–D–3094.

RED BLUFF DIVERSION DAM
PORE PRESSURE METER NOTES

OBSERVER _Wolffler_ TIME _1:00_ TO _2:45_ DATE _1/14/85_
AIR TEMP. F° _68_ RESERVOIR W.S. ELEVATION _251.13_ TAILWATER ELEVATION _238.70_

METER	SWITCH POS.	RESISTANCE OHMS	RESISTANCE RATIO	REVERSE RATIO	RESISTANCE TO GROUND	REMARKS
TEST COIL	1	50.14	.9978	1.0032		
H1	2	74.12	1.0278	.9679		
H2	3	73.04	1.0322	.9649		
H3	4	74.24	.9926	1.0124		
H4	5	73.25	1.0327	.9651		
TEST COIL	TO	H16H	TO	READ		
H5	2	74.82	1.0384	.9535		
H6	3	74.64	1.0127	.9730	GROUNDED	
H7	4	73.89	1.0322	.9548		
H8	5	74.09	1.0248	.9607		
TEST COIL						
H9	7	73.77	1.0206	.9643		
H10	8	74.58	1.0236	.9755		
H11	9	74.28	1.0202	.9674	GROUNDED	
H12	10	74.23	1.0360	.9512		

NOTES: GATE NO. 11 OPEN 2.5 FEET
         "   " 4 " .5 "
         "   " 3 " 1.5 "
         "   " 2 " 1 "
         "   " 1 " .5 "

All other gates are closed.

Figure 4-39.—Field data form for pore pressure meters. 602-D-535.

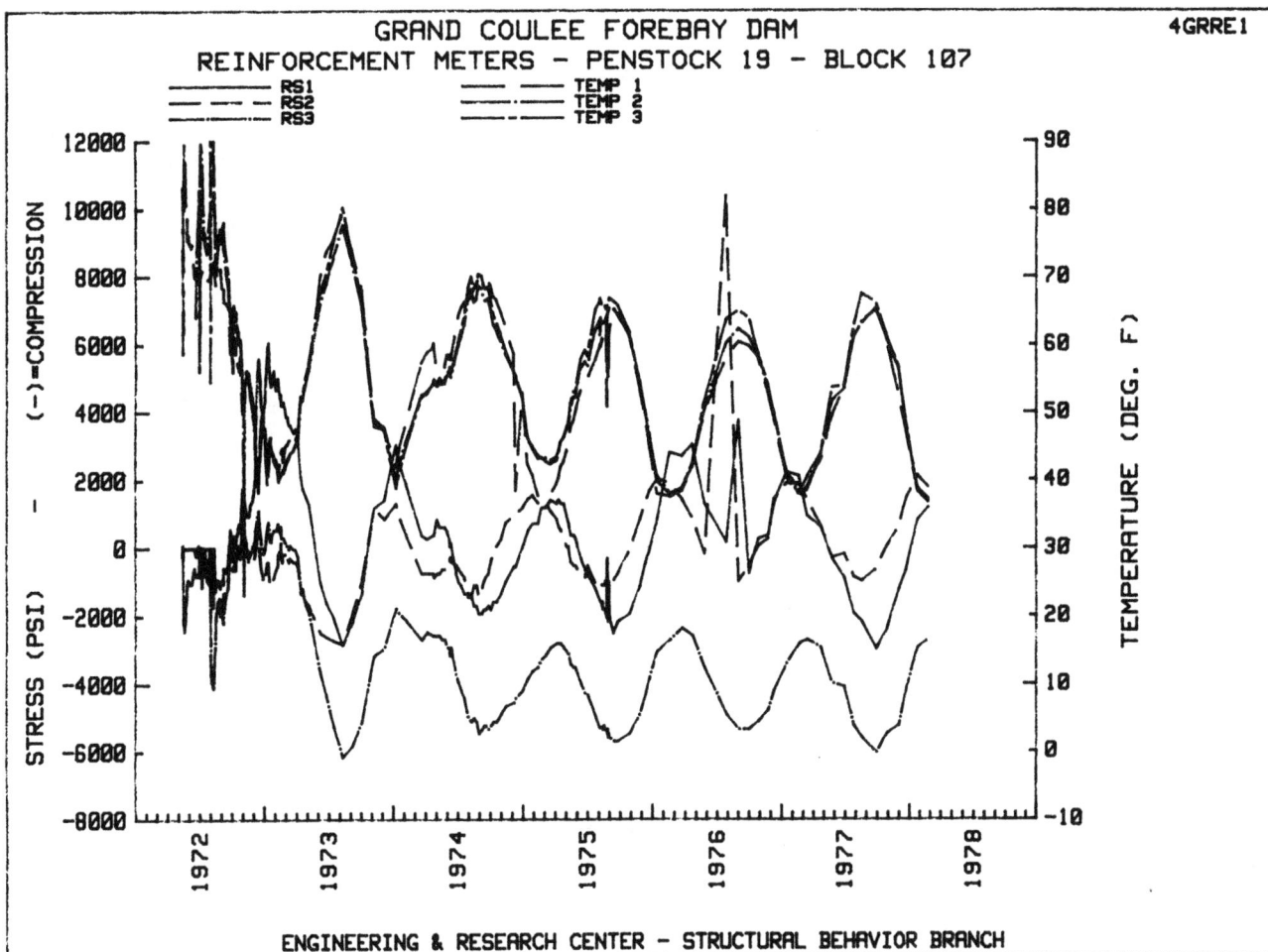

Figure 4-40.—Data plots on reinforcement meters.

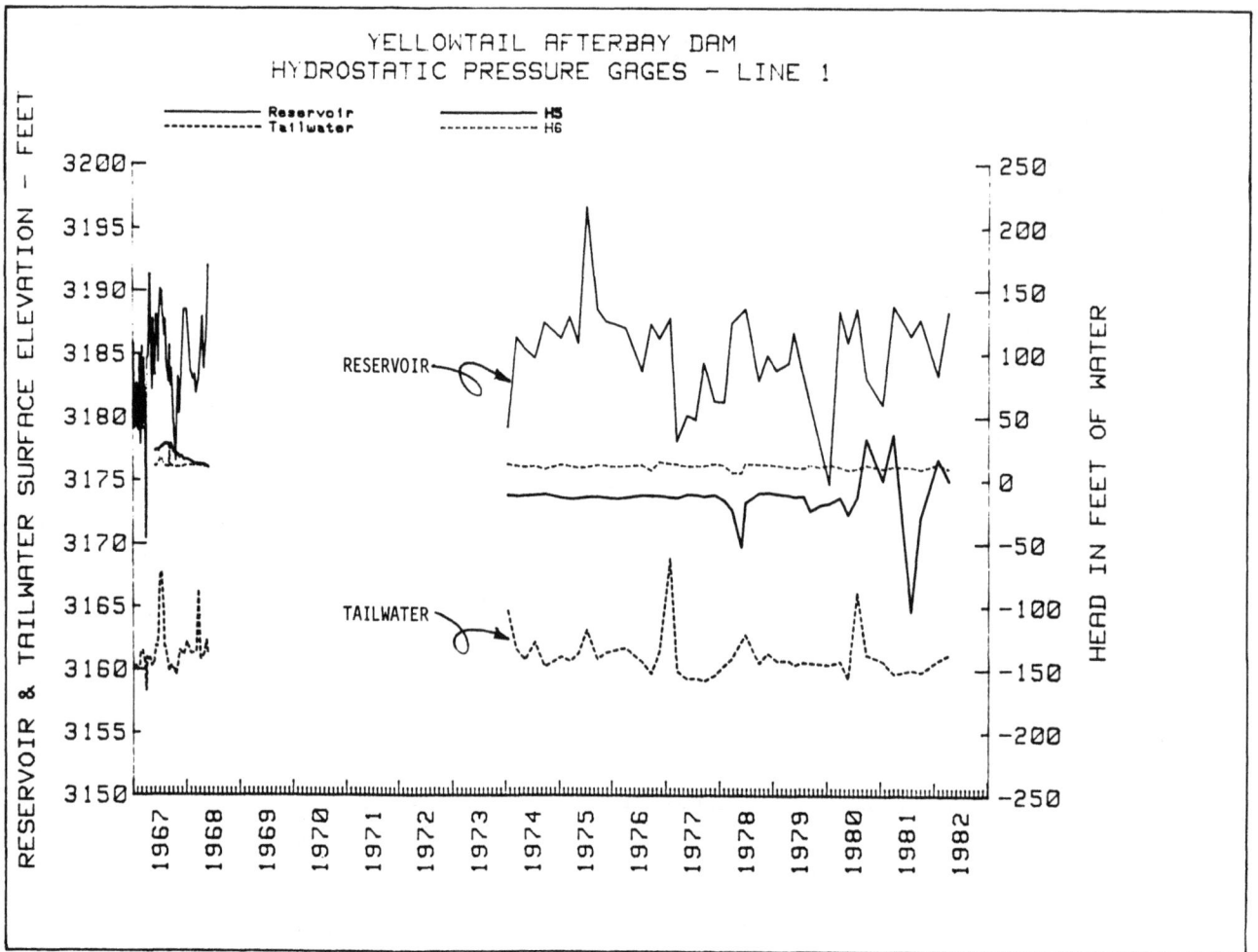

Figure 4-41.—Data plots on hydrostatic pressure gauges.

5.1 General Considerations.—Surface measurement devices generally consist of two major types: (1) those that use monuments and markers or targets to measure movements of the dam from a remote point, and (2) those that are mounted on the exterior or interior surfaces of the dam from which differential movements of portions of the structure are determined, including measurements at joints or cracks. The devices are used to measure total or relative horizontal, vertical, or rotational movements or differential movements in any desired plane. Examples of systems using monuments and markers are collimation (tangent-line measurements), levels, trilateration and triangulation, and the more common surveying angle-distance closure systems. These techniques are discussed in the following sections.

5.2 Collimation Measurements.—a. *General.*—Collimation or tangent-line measurements are a means for determining the deformation of a concrete dam with respect to references located off the dam. The bending, tilting, or horizontal displacement of a concrete dam may usually be detected by accurately measuring changes in the horizontal position of the various portions of the dam. Measurements of the location of dam monoliths at regular intervals over several years provide an indication of the magnitude of deformations that occur in the structure, its foundation, and abutments.

Alignment observations supplement previously discussed plumbline deflection measurements and the theoretical load analyses in determining the relative magnitude of the horizontal displacements resulting from structural deflection or foundation deformation. These data are valuable as an indication of the stability of the structure and also furnish information regarding the accuracy and validity of the various design assumptions and analysis procedures.

b. *Advantages and Limitations.*—These methods provide information on the magnitude and direction of movements of a dam with reference to the dam's surroundings, not merely relative movement of points on the dam. The surveying methods used are relatively common techniques; however, they must be performed to a high degree of accuracy to provide meaningful information. Instrument piers and targets located off the dam must be constructed so that they will not move and thereby destroy the accuracy of the measurements on the dam.

c. *Description of Devices.*—The instruments and devices used for collimation measurements include theodolites, stationary instrument piers and plates, moveable and fixed targets, and embedded markers or plugs.

(1) *Precision Surveying Equipment.*—All surveying readings are taken with highly precise surveying equipment. Care should be taken to achieve precise and accurate data when using theodolite equipment.

(2) *Instrument Piers and Targets.*—Instrument piers are located on one of the abutments at a concrete dam, usually the one that has the easiest access for instrument setup, and should be founded on solid rock when possible. Details of a typical installation are shown on figure 5-1. Depending upon the curvature of an arch dam and the number of points to be monitored on the crest, more than one pier may be required.

On the opposite abutment, sighting targets are installed at as many locations as required to accommodate the moveable targets located at the measurement points on the dam. These sighting targets are installed in 1.5-inch (38-mm) diameter pipes that have been solidly cemented into holes drilled into the abutment. A detail of a typical installation is shown on figure 5-2.

(3)*Moveable Targets and Plates.*—The moveable targets, which fit on plates affixed to the dam, are as shown on figure 5-3. The plates are common to the fixed piers and the dam targets. The three 120° V-slots in the anchor plate (fig. 5-3) ensure correct alignment. Measurements must consistently be taken from the same fixture post to the moveable wheel.

d. *Installation Procedures.*—As previously discussed, instrument piers installed on the abutment may be constructed during or after construction of the dam. However, it is essential that these monuments do not

PLAN

Concrete pad

5-0 × 5-0 (outer), 4-0 × 4-0 (inner)

A — A

SECTION A-A
TRILATERATION CONTROL MONUMENT
NO SCALE

3/8 16 NC x 1-0 long, brass thread stock

Brass plate (See Detail)

8 I.D. white SCH 40 PVC pipe

Sand filled void

WWF 6 x 6, W2.5 / W2.5

2 1/4 I.D. x 1-4 Long Pipe with caps

Tooled edge

10 Gates form

4:1 Sand-cement grout backfill

2 Iron pipe, concrete filled, driven into bottom of hole

8 Min. I.D. auger hole, grout filled

(dimensions: 1-0, 4-0, 3-0, 1-0, 1-0, 9-0, 12-0)

INCLINOMETER TARGET CAP DETAIL
NO SCALE

Drill and tap for 5/8 11 NC

5/8 11 NC Hex. Head bolt 1 long. Place washers under head of bolt to adjust thread length

Standard 3.38 aluminum inclinometer casing cap

Dimensions shown in inch-pound units,
1 inch = 25.4 mm
1 foot = 0.3048 m

EMBANKMENT MEASUREMENT POINT DETAIL
NO SCALE

Top must not be flared (See Notes)

15 ± Dia.

6

Riprap, topsoil or original ground

Concrete placed as to drain away from bar

1-3 ±

2-0 ±

6-0 ±

Grout in riprap only

Place bar vertically

1 O.D. Round steel bar or No. 8 rebar

3

BRASS PLATE DETAIL
NO SCALE

.25

1

6 1/4

7 3/4

5/8-11 NC bottom tapped stud hole 1/2 deep

Drill and tap 3/8-16 NC 3/4 deep, 4 req'd

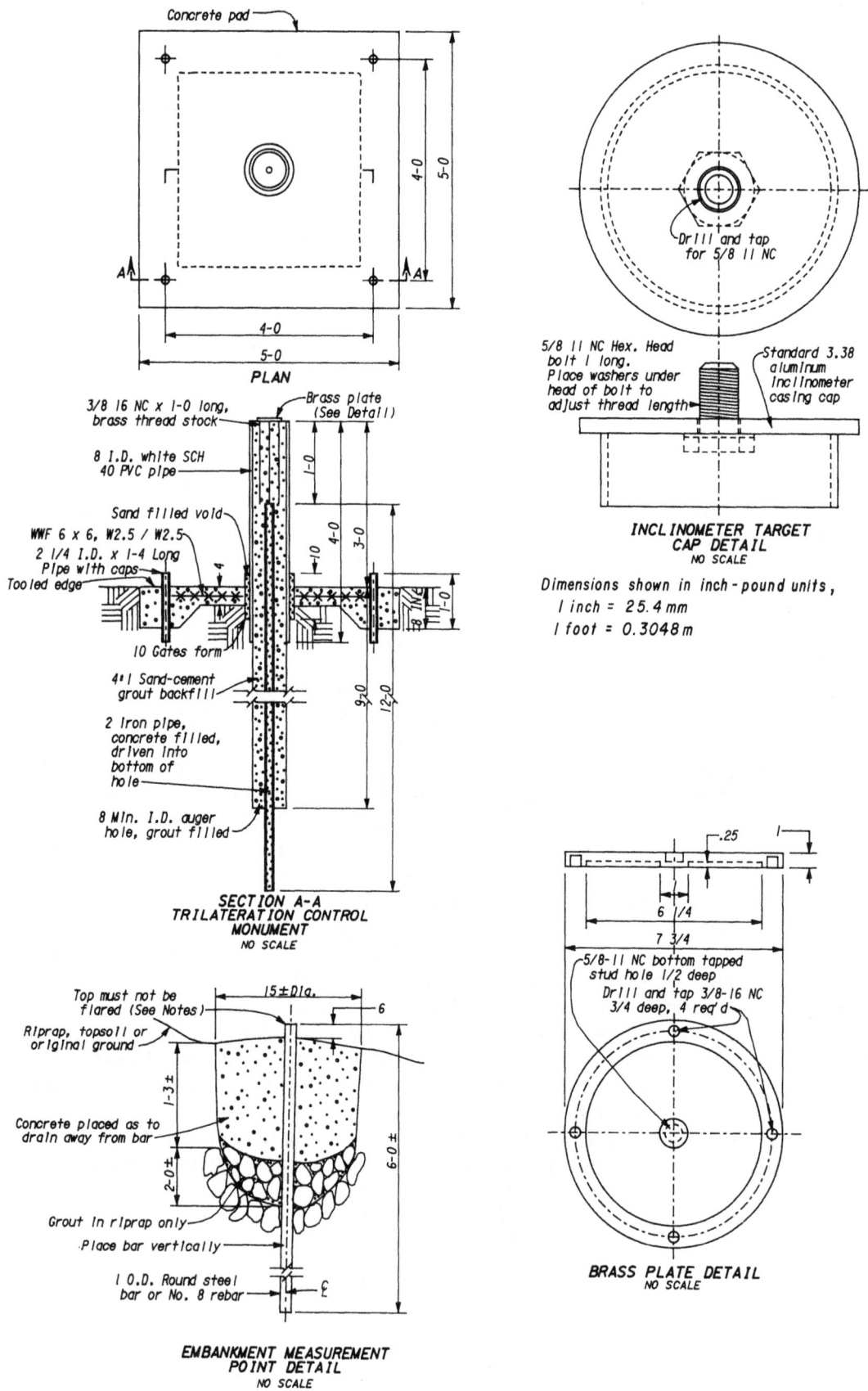

Figure 5-1.—Typical installation of an instrument pier. 963-D-125.

98

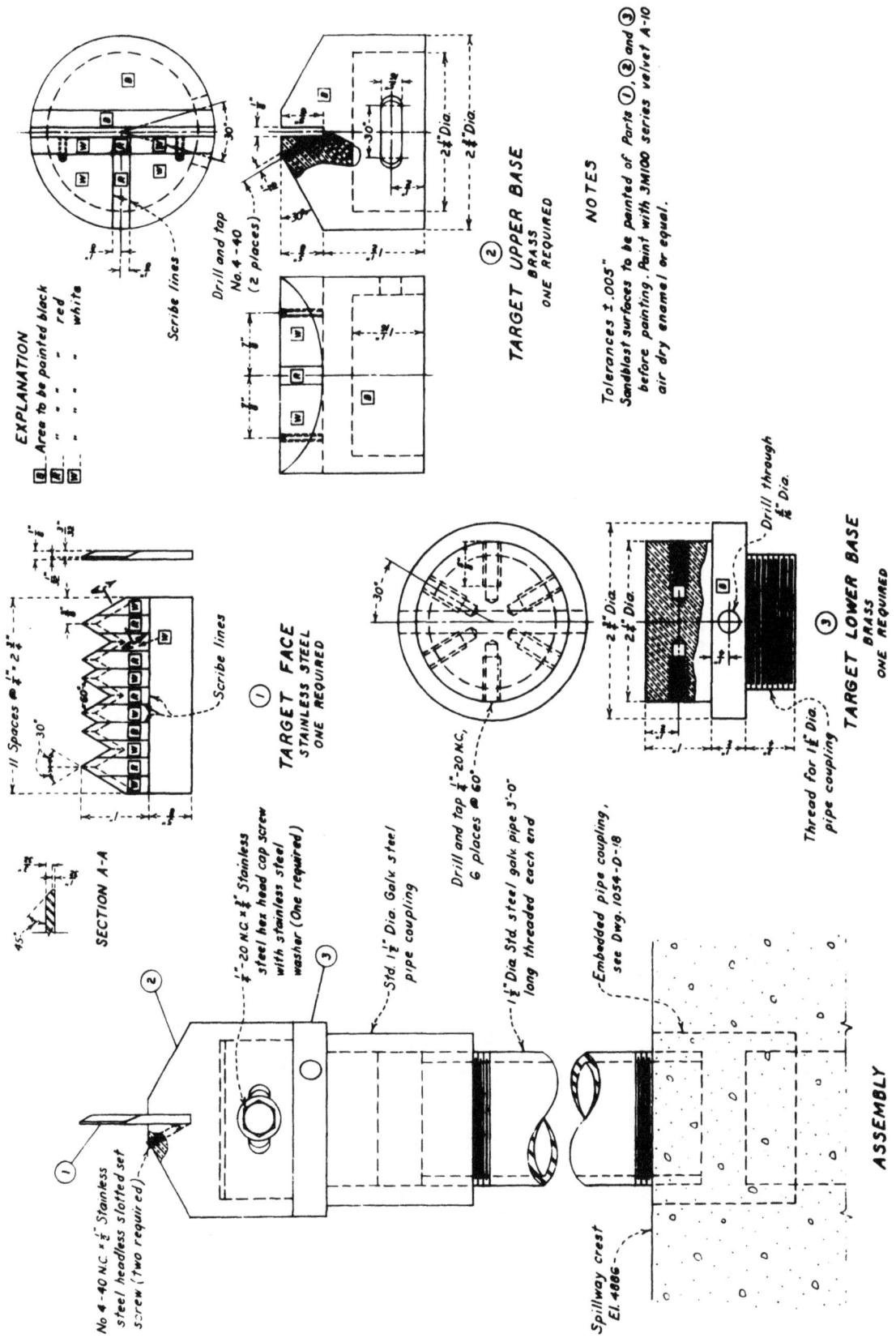

Figure 5-2.—Typical instrument target installation. 1054-D-85.

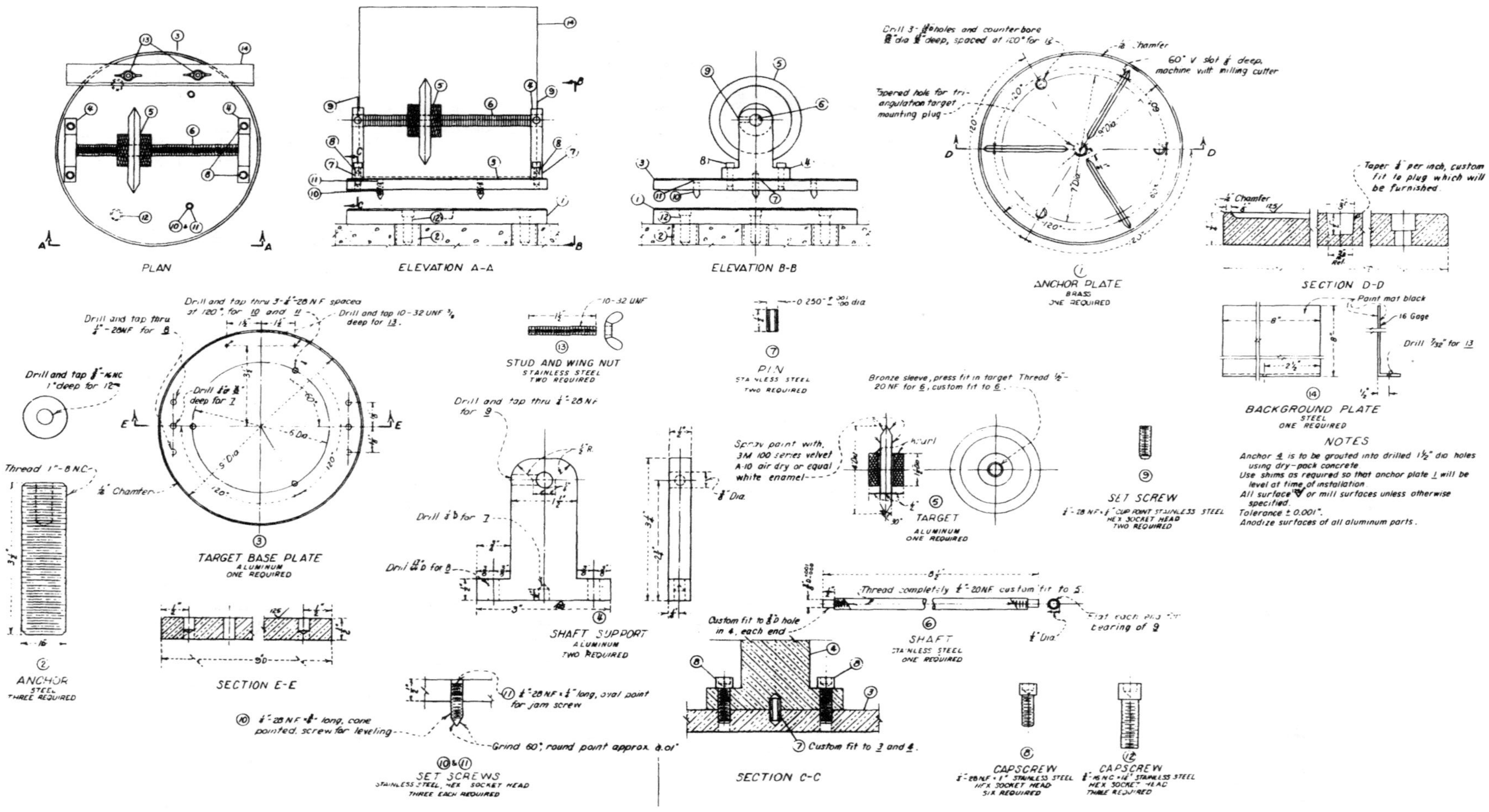

Figure 5-3.—Details of movable target assembly. 40-D-6128.

100

change position; therefore, they are constructed of concrete securely anchored in place below frost depth. Figure 5-4 illustrates a typical layout of a collimation system.

For locations on top of the dam, target plates are installed after construction has been completed. These plates are installed by drilling three holes at the proper location and grouting the anchor bolts and plates in place. Consideration should be given to the reservoir water load, season of the year, and the amount of heat remaining in the structure when locating the position of the plates so that subsequent deformation of the structure will not exceed the range or limit of movement of any collimation target.

e. Monitoring Procedures.—The measurements are made using first-order equipment, methods, and procedures insofar as feasible. The results of the measurements show deformation of a dam with respect to off-dam references. When measuring to the moveable target wheel, the wheel should be moved in the same direction each time so that any slack in the wheel will always be accounted for. Figure 5-5 shows a typical field data form.

f. Maintenance.—Maintenance of collimation systems involves keeping the measurement points clear and visible from the necessary reference lines and points. Normal maintenance of the surveying equipment is also required.

g. Data Processing and Review.—All field data forms should be transmitted to the Denver Office for processing and review as described in section 1.25. Data plots are prepared as shown on figures 5-6 and 5-7.

5.3 Trilateration Using Electronic Distance Measurements.—a. **General.**—Over the past 25 years, the development of EDM (electronic distance measurement) equipment has resulted in significant improvements in many conventional surveying methods. Locating a point or triangle by measuring distances is known as the trilateration method, and measurement techniques for this method now commonly utilize EDM devices.

The EDM device uses the velocity of electromagnetic radiation to measure distance, some types use infrared light while others use microwave radiation or visible light. The majority of the equipment currently in use

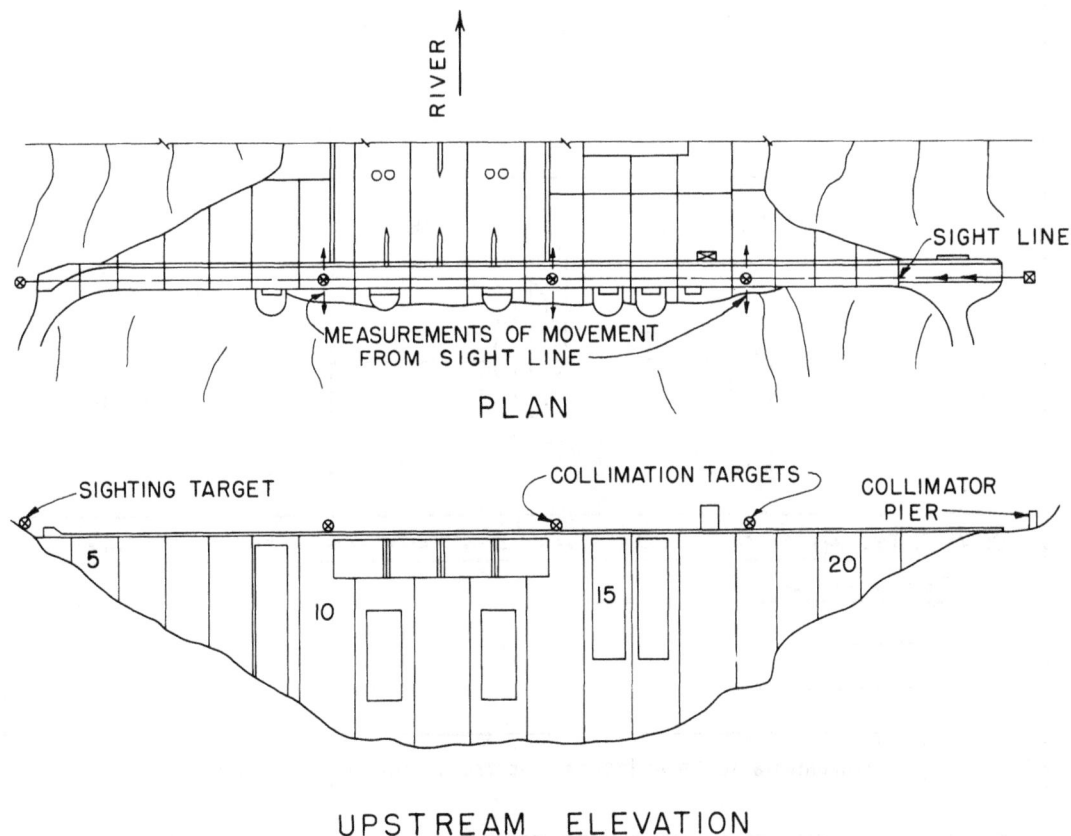

Figure 5-4.—Typical layout of a collimation system. 288-D-3097.

101

UNITED STATES
DEPARTMENT OF THE INTERIOR
BUREAU OF RECLAMATION
COLORADO RIVER STORAGE PROJECT
GUNNISON DIVISION-CURECANTI UNIT—COLORADO
CRYSTAL DAM
COLLIMATION DATA

Res. El. *6749.08* Recorder *JONES* Observer *HOBAUGH* Date *7-13-87*

BLOCK 9   Sta. 4 + 45.20

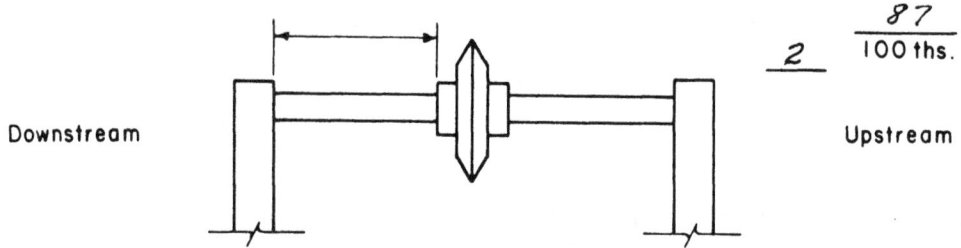

Downstream   Upstream

$2 \frac{87}{100 \text{ths.}}$

BLOCK 11   Sta. 5 + 52.48

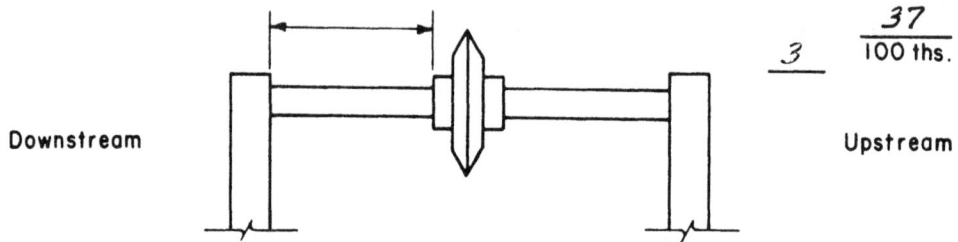

Downstream   Upstream

$3 \frac{37}{100 \text{ths.}}$

BLOCK 14   Sta. 6 + 75.29

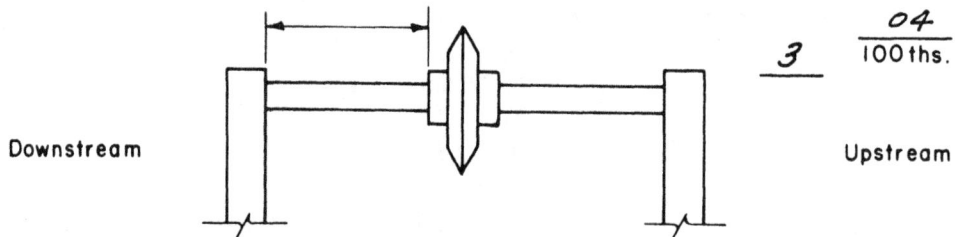

Downstream   Upstream

$3 \frac{04}{100 \text{ths.}}$

COMMENTS: *CLEAR-SLIGHT WIND BLOWING DOWNSTREAM*
0850 a.m. 57°F

Note: All measurements to be in inches plus fractional parts of inch in 100 ths.

Figure 5-5.—Field data form for collimation measurements. 622-D-3059.

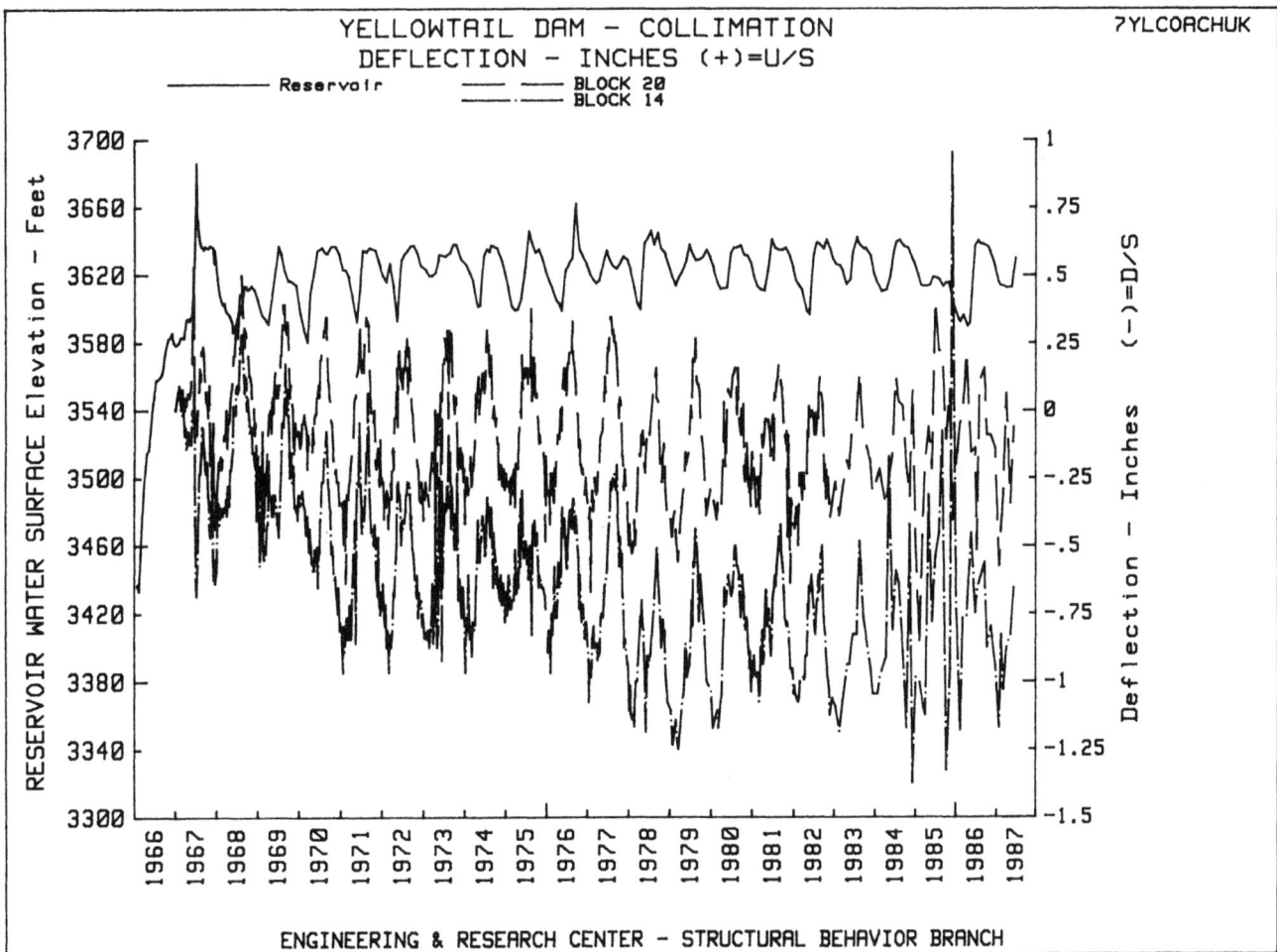

Figure 5-6.—Data plots of collimation measurements at Yellowtail Dam.

by the Bureau uses infrared light. The basic principal is to emit a modulated beam of light and observe its reflection from a target reflector. In practice, the phase angle of the reflected modulated beam is compared with the outgoing beam to arrive at a time lag. Because this method gives only the lag of multiple wave lengths, several different frequencies are used in turn, and the final distance is computed by combining the various phase relationships.

The density of the ambient air has an effect on the velocity of light; therefore, air temperature, pressure, and, to a minor extent, humidity should be measured and included in the computations. In some more recently developed EDM devices, this information is all processed internally with a microcomputer.

The EDM, depending on the type and manufacturer, can have a range from a few feet to several miles. Most of these instruments have a constant error plus an error based on a small percentage of the distance; therefore, the overall accuracy (on a percentage basis) generally increases with distance.

b. *Advantages and Limitations.*—Trilateration measurements are useful when it is not practical to measure the distance between observation points directly with tape. These measurements also permit the positioning of reference monuments at distances great enough to remove them from the danger of being involved in any general movement of the area to be monitored.

Any errors induced in a triangulation measurement method are due to distance inaccuracies rather than angular measurements. A well-trained surveying crew is an absolute necessity for accurate measurements. An example of a trilateration network is shown on figure 5-8.

c. *Description of Devices.*—The EDM devices currently used by the Bureau include devices manufactured by the Wild Heerbrugg Instrument Co., Kern and Co., Topcon Co., and the Hewlett-Packard Co. The accuracies of these devices are typically about 3 to 5 mm plus 1 to 5 p/m (parts per million), although one type has an accuracy of 0.2 mm plus 0.5 p/m.

CRYSTAL DAM
COLLIMATION – BLOCK 9 – BLOCK 11

———— Reservoir   ———— BLOCK 9
                        —·—·— BLOCK 11

ENGINEERING & RESEARCH CENTER – STRUCTURAL BEHAVIOR BRANCH

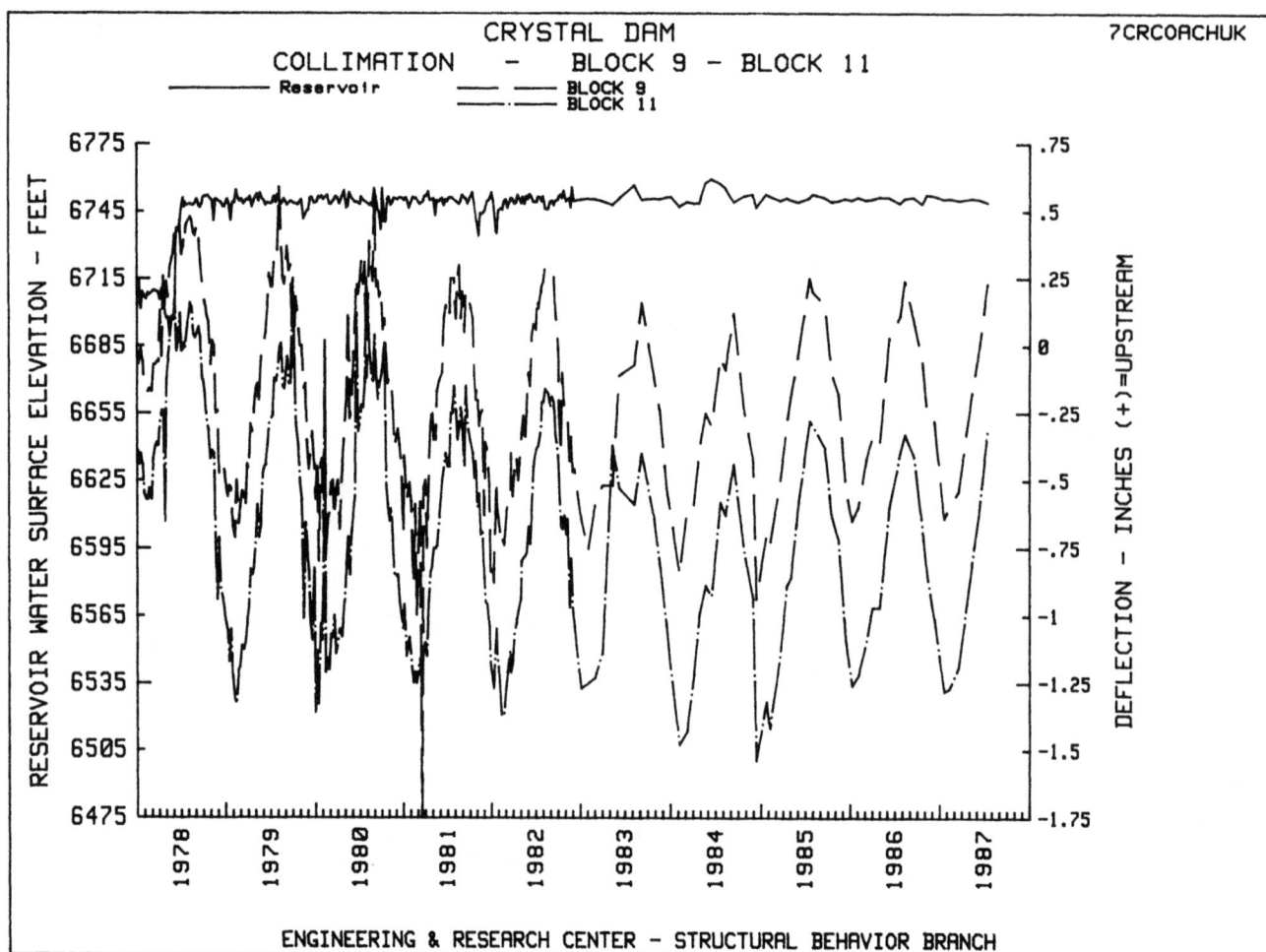

Figure 5-7.—Data plots of collimation measurements at Crystal Dam.

The reflector equipment used with EDM devices generally consists of one or a group of reflector prisms attached to a prism holder that is, in turn, mounted onto some part of the dam. Figure 5-9 shows a typical installation.

   d. *Installation Procedures.*—The installation of measurement points, observation points, and instrument piers is similar to that discussed in section 5.2.d.

   e. *Monitoring Procedures.*—Monitoring an EDM network consists of setting up the EDM device at a reading station, setting reflectors at various other stations, and taking readings. Because atmospheric conditions affect the results, the barometric pressure, temperature, and humidity should be measured, preferably along the line-of-sight. Corrections to the distances obtained are then made to allow for the atmospheric conditions. It is desirable to repeat the measurement by reversing the locations of the EDM device and the reflector; however, in some instances, accuracy is lost in instrument setups using a tripod. In general, measurements should be repeated until a consistency (first-order survey) between measurements or closure is achieved within an actual error of 1 part in 300,000. Field data are recorded on a field data form as shown on figure 5-10.

   f. *Maintenance.*—Maintenance of the measurement points and piers involves keeping the locations clear and visible from other measurement locations. Normal maintenance of the EDM equipment should be as recommended by the manufacturer.

   g. *Data Processing and Review.*—All field data forms are transmitted to the Denver Office for processing and review as described in section 1.25. Data plots are then prepared as shown on figures 5-11 and 5-12.

5.4 Triangulation.—a. *General.*—Triangulation measurements consist of measuring each of the angles in a network and then computing the distances between each measurement point. Changes in these distances

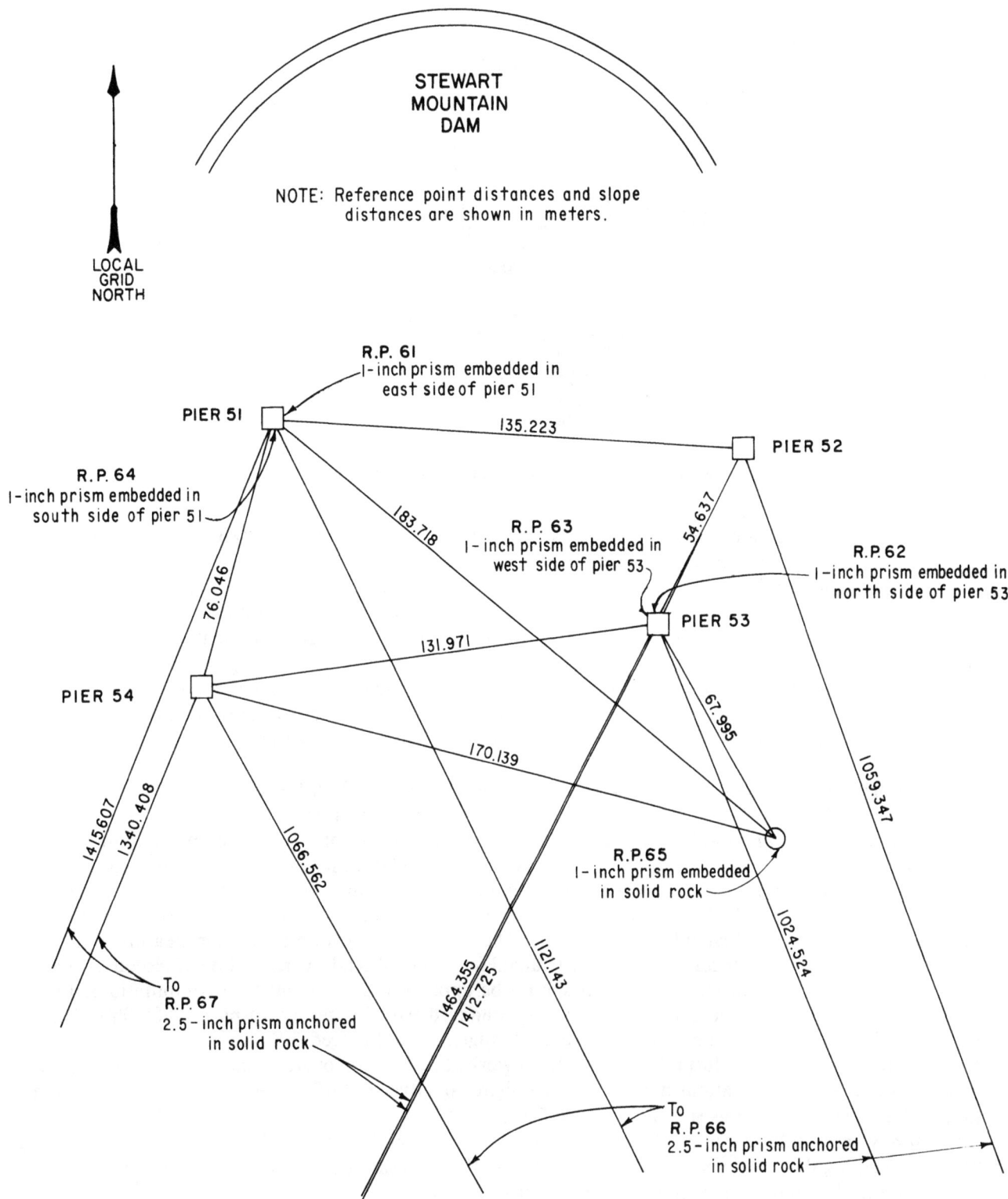

Figure 5-8.—Example of a trilateration network.

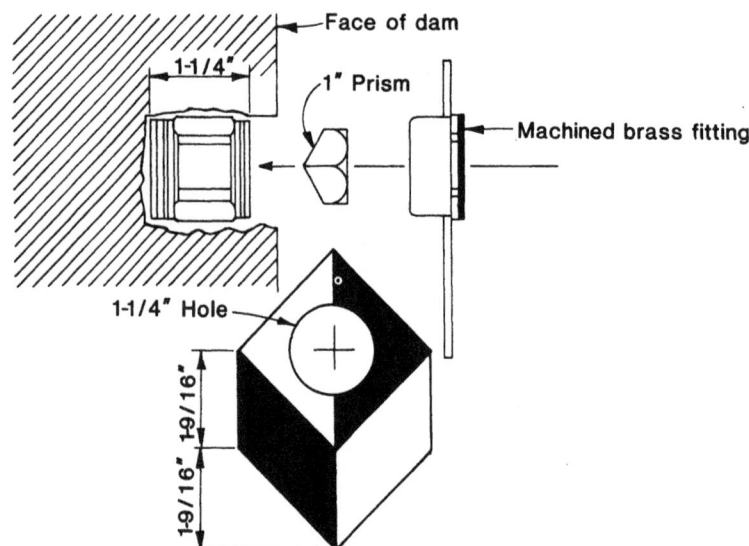

Figure 5-9.—Typical reflector equipment installation.

indicate movements of the dam. Theodolites are usually used for the angle measurements. It is suggested that a good textbook on surveying techniques be reviewed prior to all survey work.

    b. *Advantages and Limitations.*—Triangulation measurements are useful when it is not practical to measure distances directly. These measurements permit the positioning of reference monuments at distances great enough to remove them from the danger of being involved in any general movement of the area to be monitored.

Triangulation requires very precise measurements of angles and the distance along one or more base lines. Once such a system is established, it is necessary only to measure the angles to the observation points and then calculate their movements. A well-trained survey crew and accurate instruments are required to make these measurements.

    c. *Description of Devices.*—The base line or lines can be measured by EDM techniques or by conventional tape measuring methods. A theodolite is usually used for measuring the angles.

    d. *Installation Procedures.*—The installation of measurement points, observation points, and instrument piers is similar to that discussed in section 5.2.d. A typical triangulation network is shown on figure 5-13.

    e. *Monitoring Procedures.*—The angles for a triangulation system are usually measured with a 1-second theodolite. Each angle should be turned, with the instrument direct and reversed, for a total of 12 times using different positions on the graduated horizontal circle. The rejection limit for any angular measurement from the mean should be about 5 seconds, and each triangular closure should average about 2 seconds, with 5 seconds maximum. A least squares adjustment should be applied to all triangulation computations. In addition, the baseline measurements should not contain a standard error greater than 1 part in 800,000. Control points should be selected to fit the topography, and all angles should exceed 30°.

It is generally desirable to perform the triangulation work at night or on overcast days because atmospheric turbulence or temperature variations due to direct sunlight can adversely affect the accuracy of observations. Field data forms are prepared as shown on figure 5-14.

    f. *Maintenance.*—Maintenance of the measurement points and piers involves protecting them from damage and keeping the locations clear and visible from other measurement locations. Normal maintenance of the theodolite should be as recommended by the manufacturer.

    g. *Data Processing and Review.*—All field data forms are transmitted to the Denver Office for processing and review as described in section 1.25. Data plots are then prepared as shown on figure 5-15.

    5.5 Precision Leveling.—a. *General.*—Vertical movements of concrete dams are usually measured by precision leveling techniques from known benchmark elevations to various points on the dam. When such observations are made at relatively regular time intervals, the rate of movement and the total movement can be determined.

UNITED STATES
DEPARTMENT OF THE INTERIOR
BUREAU OF RECLAMATION
CENTRAL VALLEY PROJECT-CALIFORNIA

MONTICELLO DAM

HORIZONTAL DISTANCES USING KERN MEKOMETER-3000

RESERVOIR ELEV. 424.45 OBSERVER Stackhouse AIR TEMP. 65 °F
RECORDER Lorenz DATE 6-5-87

RT. PIER		LT. PIER	
LT. PIER	374.2732 m	RT. PIER	374.2729 m
PIER B-3	680.3030	PIER B-3	660.6329
PIER B-4	711.6277	PIER B-4	593.5190
NO. 3A	769.1108	NO. 5	669.6585
NO. 3	772.2041	NO. 7	691.5916
NO. 5	785.8236	NO. 9	708.8427
NO. 7	793.8576	NO. 11	720.9215
NO. 9	796.1443	NO. 13	727.5318
NO. 11	792.6278	NO. 15	728.4821
NO. 13	783.4231	NO. 17	723.7180
NO. 15	768.7505	NO. 19	714.3980
NO. 17	748.9975	NO. 20	710.1443

PIER B-3		PIER B-4	
PIER B-4	177.7826	PIER B-3	177.7834
LT. PIER	660.6356	LT. PIER	593.5210
RT. PIER	680.3055	RT. PIER	711.6308

COMMENTS:

Figure 5-10.—Field data form for a trilateration system. 413-D-628.

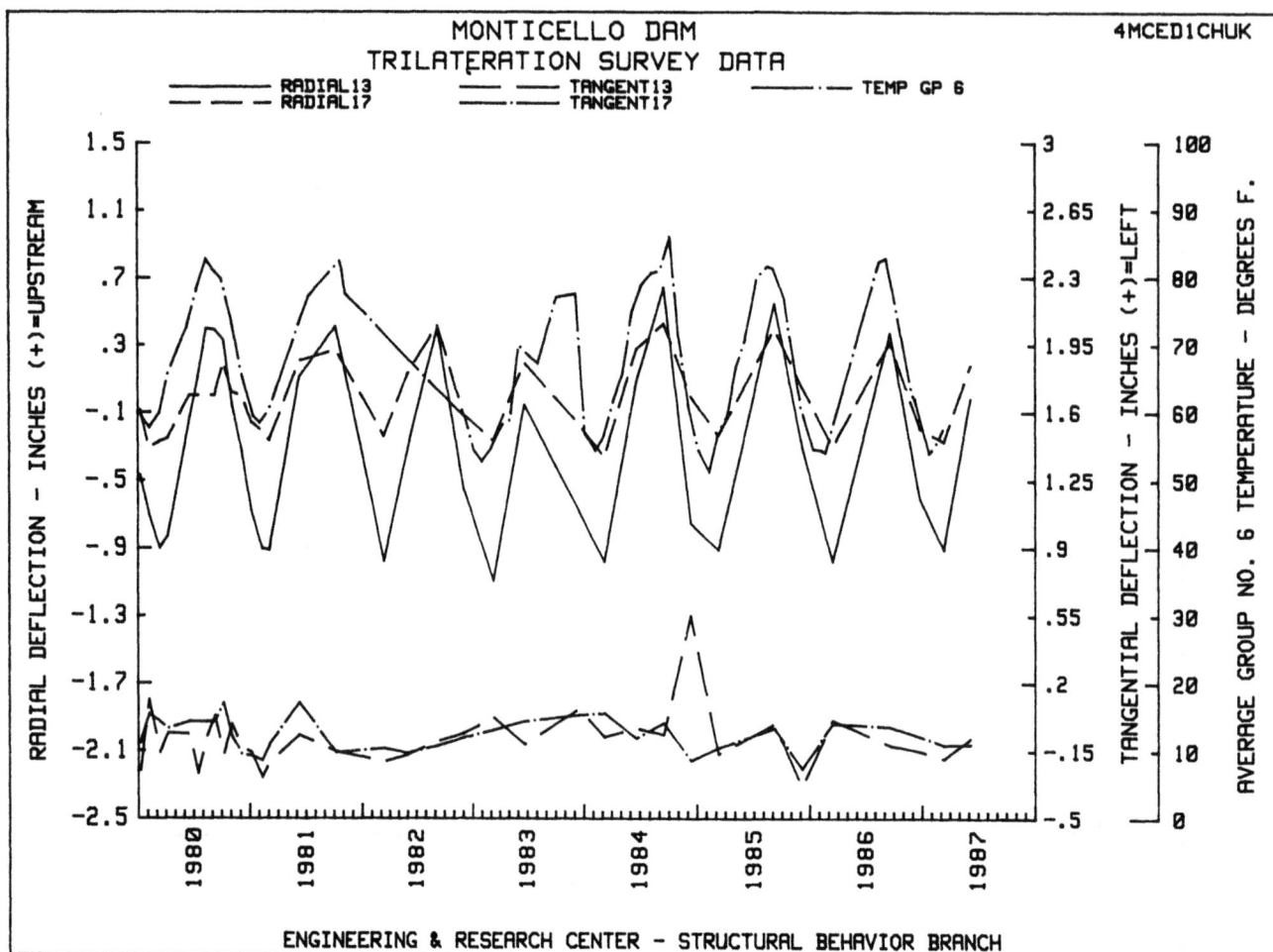

Figure 5-11.—Data plots on trilateration survey at Monticello Dam.

b. *Advantages and Limitations.*—Leveling techniques are virtually the only way that vertical movements of a dam relative to off-dam benchmarks can be efficiently detected and measured. To obtain the proper accuracy, the engineer must recognize that some errors will be involved, and an effort to minimize those errors must be made. For example, precision leveling equipment used by an inexperienced crew may possibly result in a greater error than that from less precise equipment used by an experienced crew.

In general, the equipment used should be of an accuracy higher than that desired for the final result. Most reputable manufacturers include a statement of accuracy in their specifications for a particular instrument that is reliable if operated in accordance with the instructions.

c. *Description of Devices.*—The leveling devices used vary from conventional engineer's levels to precise levels that include tilting levels, self-leveling devices, and optical-tooling levels. Their precision is approximately as follows:

Device	*Precision*
Conventional Levels	±0.03 feet (±0.009 m) in 3,280 feet (1 km)
Precision Levels:	
Tilting	±0.003 feet (±0.0009 m) in 3,280 feet
Self-leveling	±0.007 feet (±0.0021 m) in 3,280 feet
Optical-tooling	±0.0002 feet (0.00006 m) in 100 feet (30.48 m)

The optical-tooling levels are used only for shorter distances, and have not been widely used in Bureau applications.

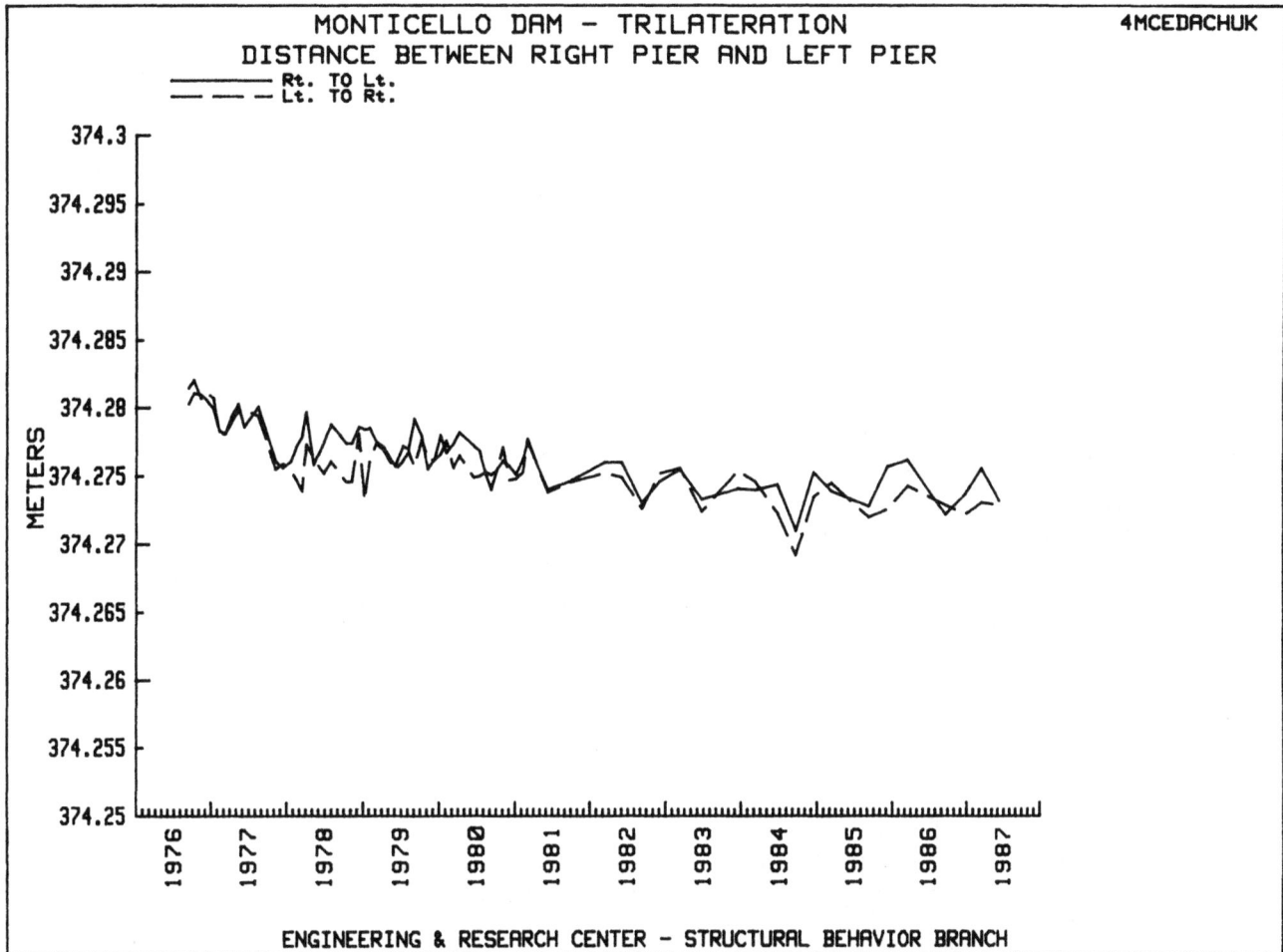

Figure 5-12.—Data plots on distances between right and left piers at Monticello Dam.

The accuracy of leveling instruments is primarily determined by the sensitivity of the level bubble, which is rated in terms of arc-seconds per 2 mm of level vial. Most engineer's levels have a bubble rating of 20 to 30 arc-seconds per 2 mm, while precise tilting levels range from 2 to 10 arc-seconds per 2 mm. The level vial sensitivity should be the limiting factor for level accuracy, and values on the accuracy of the optical train or stability of the instrument should be equal to or greater than the bubble sensitivity.

If level sights are taken over long distances, corrections must be made for refraction and earth curvature. Also, the distances for backsights to a benchmark and foresights to an observation point should be kept nearly equal. Sightings on a benchmark or an observation point may be made by reading a level rod placed on the mark or by reading a combination moveable target placed on a calibrated rod. The latter method usually yields greater precision.

d. *Installation Procedures.*—Observation points and benchmarks should be installed to a degree of fixity such that observation points move precisely with the portion of the dam where installed, and benchmarks should be located off the dam in an area which is unaffected by dam movements or by movements of the surrounding soil or rock mass. The actual installation procedure is similar to that discussed in section 5.2.d.

e. *Monitoring Procedures.*—Leveling is performed by setting up the level and reading the vertical distance to a benchmark (backsight) and then reading the vertical distance to an observation point (foresight). It is very important to try to balance the distances for the backsight and foresight to minimize any error introduced by the line-of-sight not being horizontal. After establishing the elevation of the observation point, the instrument is moved and the observation point becomes the backsight and another observation point becomes the foresight. This procedure is continued until a benchmark becomes the foresight. At that time, the computed elevation for the benchmark should check within 0.017 foot (0.0052 m) times the square root of the number

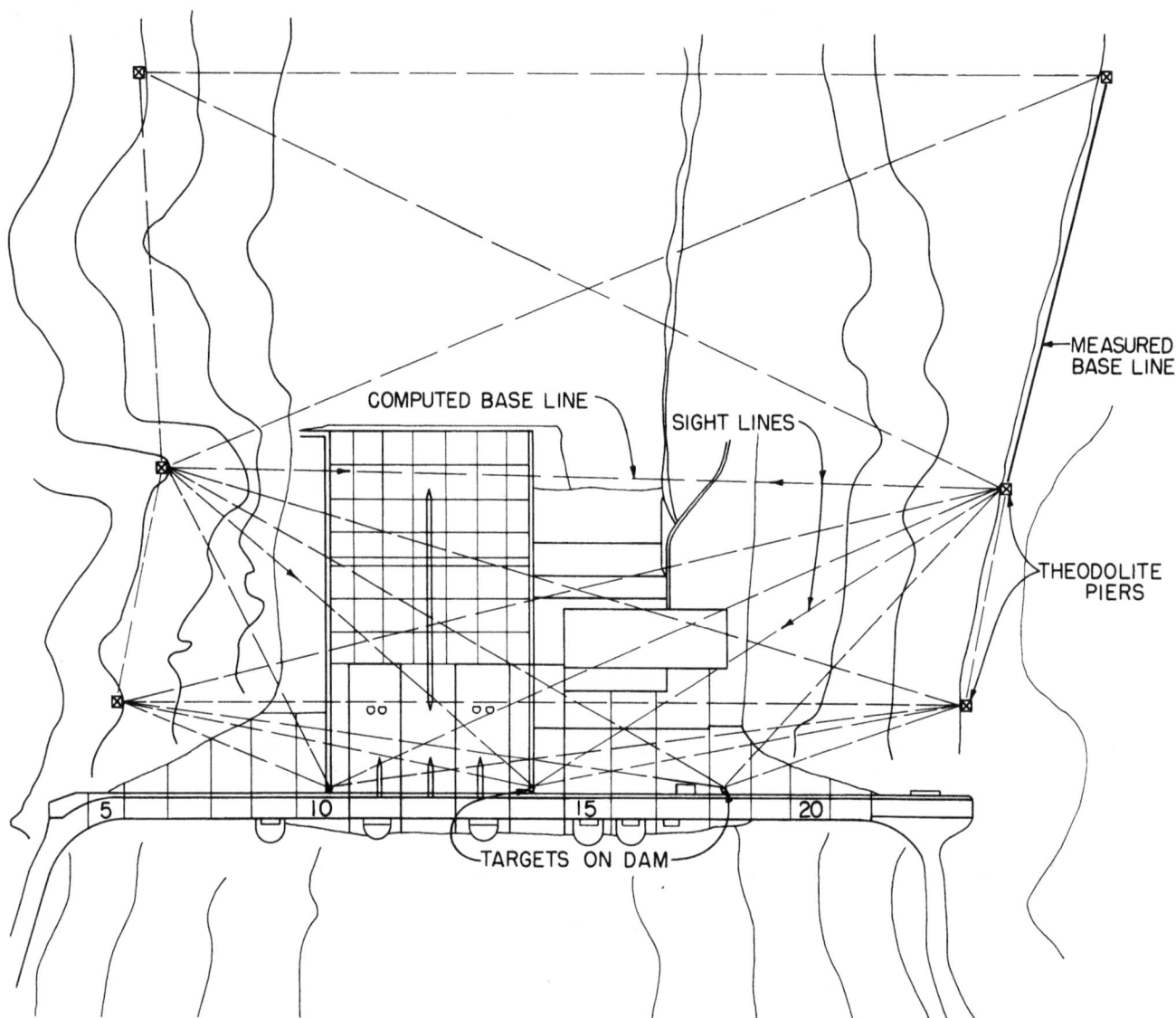

Figure 5-13.—Typical triangulation network. 288-D-3099.

of miles (kilometers) traversed of the previously known benchmark elevation, or the procedure should be repeated. A typical field data form is shown on figure 5-16.

   f. *Maintenance.*—Maintenance of the observation points and benchmarks involves keeping the locations clear so that the line-of-sight is maintained. Normal maintenance of the leveling devices should be as recommended by the manufacturer.

   g. *Data Processing and Review.*—All field data forms should be transmitted to the Denver Office for processing and review as described in section 1.25. Data plots are then prepared as shown on figure 5-17.

   5.6 Surface Measuring Points.—a. *General.*—Surface measurements are usually conducted across contraction and/or construction joints bounding the blocks containing other instrument clusters. These surface measurements measure movements transverse and parallel to the joint; however, a third dimension may be measured using special devices. These instruments provide a means for monitoring the behavior of joints (or cracks) to determine the beginning or extent of a joint opening due to cooling of the mass concrete or to applied pressures on the dam. The instruments may also serve as indicators of maximum joint opening to indicate when grouting should be performed, may indicate the effects of grouting, and show whether any movement in the joint occurs during or after grouting if the instruments were installed early in the life of a dam.

110

POWER OPERATIONS OFFICE
MONTROSE, COLORADO
(3-89)

UNITED STATES
DEPARTMENT OF THE INTERIOR
BUREAU OF RECLAMATION

SHEET NO. 3 OF 10

GROUP

COLORADO RIVER STORAGE PROJECT- GLEN CANYON DAM
STRUCTURAL BEHAVIOR STUDIES
TRIANGULATION

BY G DATE 19. Aug. 85
CHKD. BY _____ DATE _____

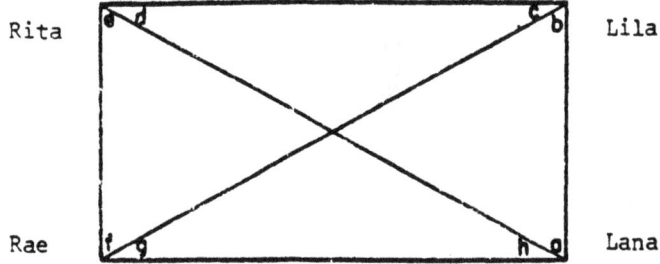

a	54	49	37.58	a
b	53	00	27.69	b
c	35	44	51.22	c
d	36	24 25	03.77	d
e	56	17	14.91	e
f	51	32	50.04	f
g	34	40	31.71	g
h	37	29	23.13	h

a	54	49	37.58	e	56	17	14.81
b	53	00	27.69	f	51	32	50.04
c	35	44	51.22	g	34	40	31.71
d	36	25	03.77	h	37	29	23.13

180 - 00 - 00.26   179 - 59 - 59.69

g	34	40	31.71	c	35	44	51.22
h	37	29	23.13	d	36	25	03.77
a	54	49	37.58	e	56	17	14.81
b	53	00	27.69	f	51	32	50.04

180 - 00 - 00.11   179 - 59 - 59.84

Figure 5-14.—Field data form for a triangulation network.

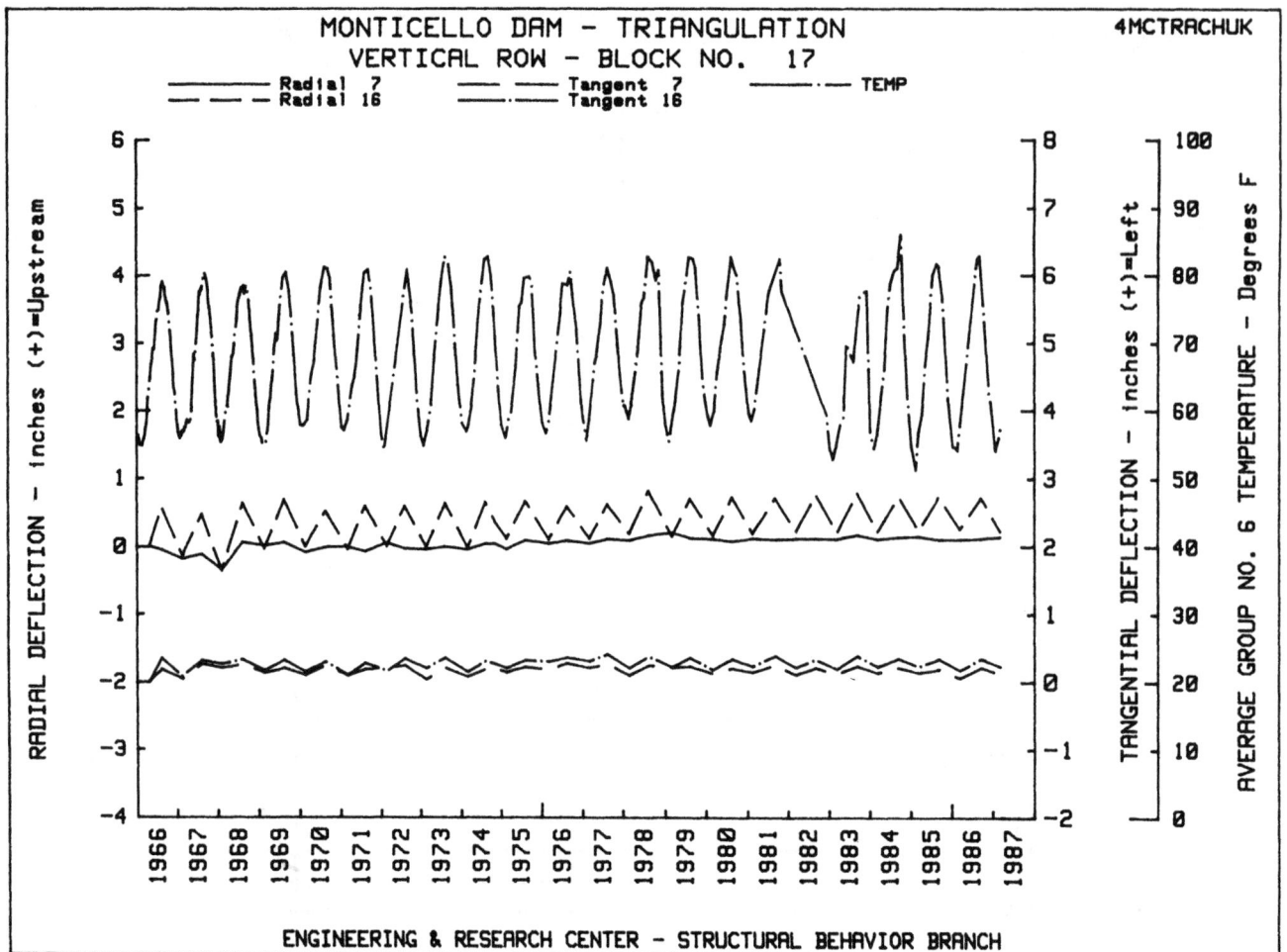

Figure 5-15.—Data plots on a triangulation network at Monticello Dam. (Sheet 1 of 2).

The instruments consist of measuring points embedded in the surface of the concrete on both sides of a joint or crack between which measurements are made at various time intervals. The measurements may be made by tapes, Whittemore gauges, dial gauges, depth micrometers, or commercial crack measuring devices. The measurement points may be installed in the top or side surfaces of a concrete dam or in the interior surfaces of galleries or tunnels.

   b. *Advantages and Limitations.*—Installation of these devices allows measurement of relative movement of adjacent blocks or monoliths of a dam and, when done in conjunction with the alignment measurements described in sections 5.2 through 5.5, the total movement of blocks can be determined. Obvious limitations include the necessity for accurate observational techniques and the necessity of obtaining precise initial position measurements.

   c. *Description of Devices.*—The measuring points may consist of virtually any kind of mark or point set in the concrete from which measurements may be made. However, it is most common to use specially fabricated brass or stainless steel carriage bolts or cap screws for certain devices such as Whittemore gauges or tapes. Figure 5-18 shows a typical installation of these measuring points.

Tapes or ruling devices used to measure the distances between the points must be of sufficient precision to allow measurement to the desired accuracy. Ames dials or Whittemore gauges (fig. 5-19) may be used for these measurements. With these devices, sufficient precision to allow observation to the nearest 0.001 inch (0.025 mm) is possible. Three points, one on one side of the joint and two on the other side of the joint, may be used to monitor opening and shear movement in the reference plane along the joint.

A patented, relatively accurate, crack measuring device known as the "Avongard Calibrated Crack Monitor" (fig. 5-20) also indicates movement in two directions.

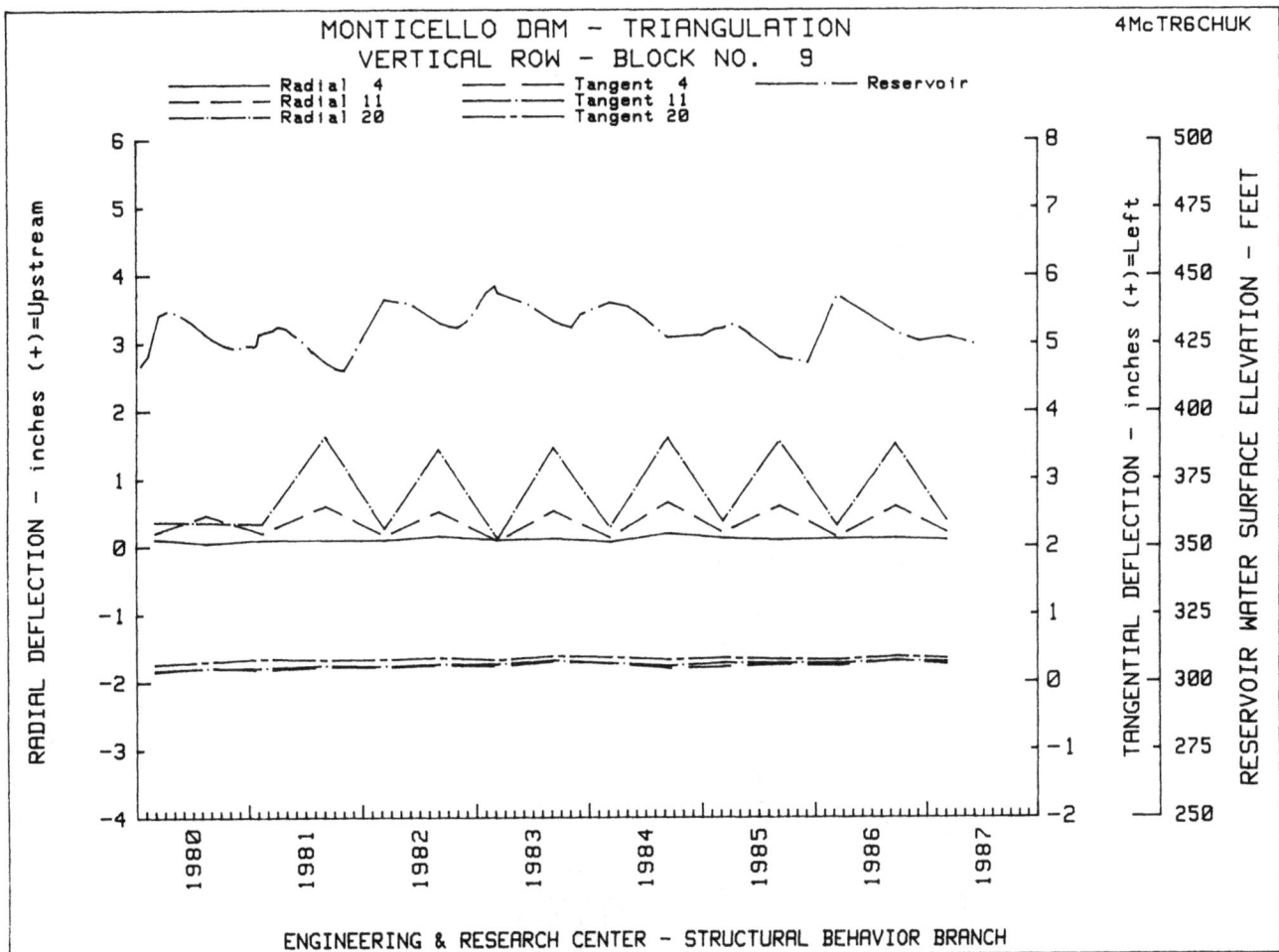

Figure 5-15.—Data plots on a triangulation network at Monticello Dam. (Sheet 2 of 2).

d. Installation Procedures.—Although simple scratch marks may be used, it is more common to use metal points embedded in the concrete. The points may be set in fresh concrete, grouted into place in drilled holes, or installed with expansion anchors after the concrete has set. For setting the ⅜-inch (9.5-mm) diameter carriage bolts, ½-inch (13-mm) diameter holes are used. The points should be set at locations where movement may reasonably be expected, may indicate a potential problem, or may result in a critical situation. The locations should be readily accessible and a clear line must be present to enable measurement between at least two points on opposite sides of the joint or crack.

The Avongard device is affixed across a crack or joint using epoxy to securely fasten each end of the device on opposite sides of the crack or joint.

e. Monitoring Procedures.—Monitoring of all but the self-reading devices is accomplished by accurately measuring the distance between the two or more points located on opposite sides of the joint or crack. Measurements are made to the desired accuracy, usually to the nearest 0.001 inch (0.025 mm). Self-reading devices may be read and recorded including the date and location, or they may be photographed with the photograph properly labeled with date and location of the device. All field data are recorded on forms such as the form shown on figure 5-21.

f. Maintenance.—The only maintenance required consists of keeping the measuring points clean and proper care of the tapes, dials, and gauges. A standard Invar reference bar should be provided for calibration of the Whittemore gauges, which should be done each time a Whittemore gauge is used.

g. Data Processing and Review.—The field data forms are transmitted to the Denver Office for processing and review as described in section 1.25. Data plots are then prepared as shown on figure 5-22.

MEASUREMENT POINTS
CUMULATIVE SETTLEMENT
AND
DEFLECTION READINGS
EMBANKMENT

TO: DIRECTOR OF DESIGN & CONST.
ENGRG & RESEARCH CENTER
P.O. BOX 25007 DEN. FED. CNTR
DENVER, COLO. 80225
ATTENTION: 220

MODIFIED................FOR

Dam...... Parker ....

Project.... L.C.D.P.O. ....

Ref. Dwg .... See Sheet 3 of 3

Date of Observation.......... 3/87

Observer........ D. Keranen ....

Sheet ..... 2 ..... of 3

| CENTERLINE STATION | SETTLEMENT | | | OFFSET FROM CENTERLINE | DEFLECTION | | |
	ORIG. EL.	PRES. EL.	CUM. DIFF.	DATE POINTS SET OR RESET	(1) ORIG. OFFSET	PRES. OFFSET	CUM. DIFF.
Compressor Room 16	391.015	391.015	.001				
17	391.374	391.382	-.008				
Pipe Gallery 18	373.342	373.344	-.002				
19	362.072	362.094	-.022				
20	362.078	362.102	-.024				
21	362.071	362.100	-.029				
22	362.059	362.085	-.026				
23	362.091	362.116	-.025				
Unwatering Gallery 24	362.122	362.148	-.026				
25	362.163	362.184	-.021				
A	362.075	362.161	-.086				
B	362.076	362.161	-.085				
C	362.094	362.178	-.084				
D	362.051	362.107	-.056				
E	362.030	362.078	-.048				
F	362.073	362.174	-.101				
G	362.083	362.172	-.089				
H	362.061	362.156	-.091				
J	362.043	362.096	-.053				
K	362.082	362.154	-.072				
Crack Survey *12/6/62 #1 N/S	2.457 *	2.441		362 Elevation			
#2 N/S	6.229 *	6.137		362 Elevation			
#3 N/S	6.343 *	6.293		373 Elevation			
#1 E/W	2.297 *	2.320		362 Elevation			
1 Vert.	2.656 *	2.636		362 Elevation			
BM 1		468.920					
BM 2	468.925	468.934	-.009				
BM 3	468.912	468.921	-.009				
BM 4	468.948	468.960	-.012				
Power Plant Roof. BM 5	468.909	468.919	-.010				
BM 6	468.903	468.910	-.007				
BM 7	468.956	468.982	-.026				
BM 8	468.959	468.988	-.029				
BM 9	468.943	468.962	-.019				
BM 10	468.949	468.962	-.013				
# 2		.01					
# 3		.01					
Power Plant Roof Alinement A	0.910	.875					
B	0.940	.932					
C	0.910	.917					
D	0.940	.938					
E	0.940	.938					
F	0.930	.922					
# 4		.01					

(1) Record u/s or d/s from ₵ Crest or Axis of Dam.
Report original elevations and offsets to 0.01 foot.
Report cumulative settlement and deflections to 0.01 foot.

Use Minus (-) sign to report heave.
Use u/s or d/s to report horizontal movement.

GPO 857-765

Figure 5-16.—Field data form for level readings.

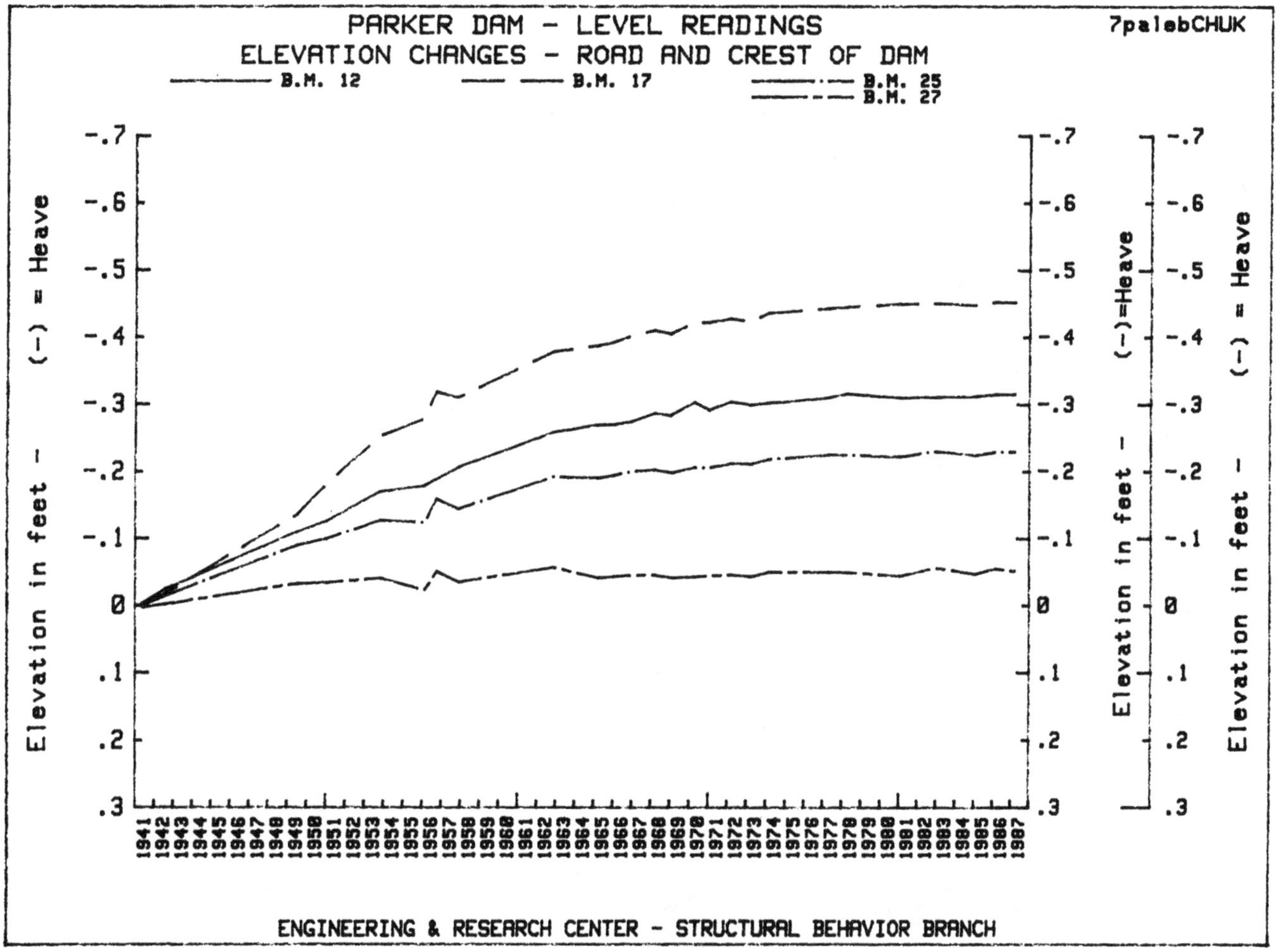

━━━━━━ B.M. 12   ━━ ━━ B.M. 17   ━━━━━·━━ B.M. 25
                                              ━━━━━·─·─ B.M. 27

ENGINEERING & RESEARCH CENTER - STRUCTURAL BEHAVIOR BRANCH

Figure 5-17.—Data plots on level readings.

5.7 Tiltmeters.—a. *Usage.*—Tiltmeters may be either portable or fixed in place, and they are rapid, easy-reading devices used to monitor the vertical tilt (rotational movement) of dams, sections of dams, and rock masses. These devices are generally affixed to the surface of the structure so that the tiltmeter plate moves with the structure. A portable tiltmeter has only its plate affixed to the structure.

b. *Advantages and Limitations.*—Tiltmeters are lightweight, compact, and are a relatively economical way of obtaining rotational movement data. However, because the sensor plate is affixed in only one spot on a structure, the overall movement of a portion of a structure can only be determined by using a number of sensor plates.

c. *Description of Devices.*—The three basic parts of a portable tiltmeter device are a ceramic plate, sensor device, and a readout indicator. Figure 5-23 illustrates this device.

(1) *Ceramic Tilt Plates.*—These are ceramic plates, cast from specially formulated porcelain, containing four sensor orientation pegs in the upper surface. The plates are typically of 6 inch (152 mm) diameter and ¾ inch (19 mm) thick, with the four pegs spaced 4 inches (102 mm) apart across the center of the plate. The ceramic material is used to ensure maximum temperature stability.

(2) *Tiltmeter Sensor.*—The portable tiltmeter sensor uses two closed-loop, force-balanced, servoaccelerometers specifically designed for tilt measurements. The accelerometers are set at 90° with respect to each other. The sensor body consists of a steel plate base and aluminum housing around the accelerometer.

(3) *Readout Indicator.*—A readout indicator is used to indicate the angle of inclination of the tiltmeter sensor. Typically, the indicator is able to read over a standard operating range of ±30° from the reference plane. The unit is powered by an internal rechargeable battery.

d. *Installation Procedures.*—The ceramic plates are installed using either a grout or epoxy resin to firmly affix the plates to the side or top of the structure or rock surface to be measured. The surface must be

Figure 5-18.—Typical installation of measuring points. P801-D-81192.

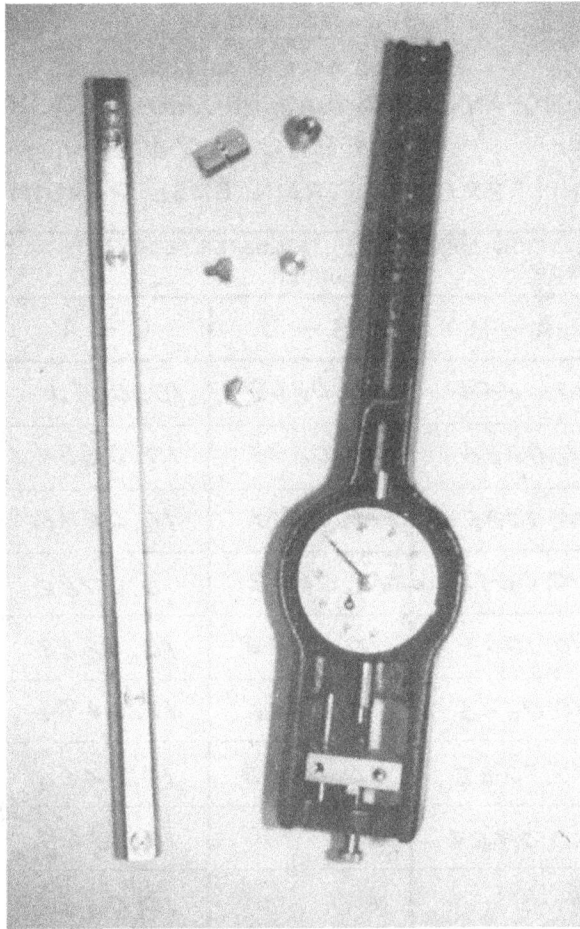

Figure 5-19.—Soiltest Whittemore gauge. P801-D-81193.

Figure 5-20.—Avongard calibrated crack monitor. P801-D-81194.

117

UNITED STATES
DEPARTMENT OF THE INTERIOR
BUREAU OF RECLAMATION
FRYINGPAN – ARKANSAS PROJECT – COLORADO
PUEBLO DAM
WHITTEMORE STRAIN GAGE READINGS

RESERVOIR W.S. ELEVATION _4806.59_ TAILWATER ELEVATION _4741.3_

OBSERVER _M. USGROVE_ RECORDER _____ DATE _12-7-79_

JOINT	A – B	B – C	C – A	STANDARD
1 – 2	10.0254	10.0352	10.0143	10.0530
2 – 3	10.0286	10.0295	10.0634	10.0532
3 – 4	10.0035	10.0541	10.0698	10.0533
4 – 5	10.0641	10.0832	10.0782	10.0537
5 – 6	10.1003	10.0728	10.0288	10.0538
6 – 7	10.0645	10.0746	10.0494	10.0539
7 – 8	10.1022	10.0868	10.0466	10.0538
8 – 9	10.0964	*	10.0845	10.0540
9 –10	*	*	10.0666	10.0538
10–11	10.0955	10.0*	10.0368	10.0537
11–12	10.0810	10.0750	10.0653	10.0536
12–13	10.0554	10.1004	10.0305	10.0537

REMARKS: _*GAUGE WOULD NOT READ_

_130.6972/13 = 10.0536_   _10.0460-10.0536 = -0.0076 (ADJ.)_

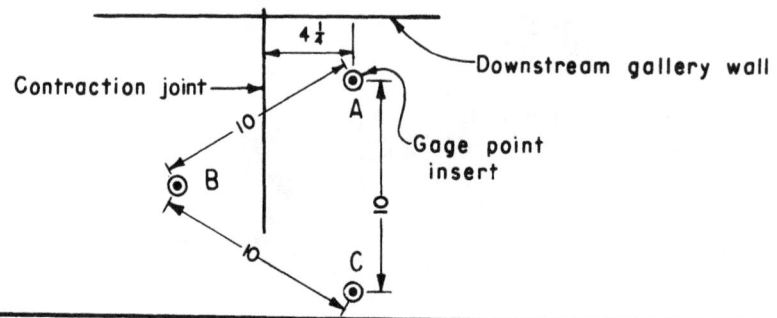

Figure 5-21.—Field data form for crack or joint monitoring. 382-D-1894.

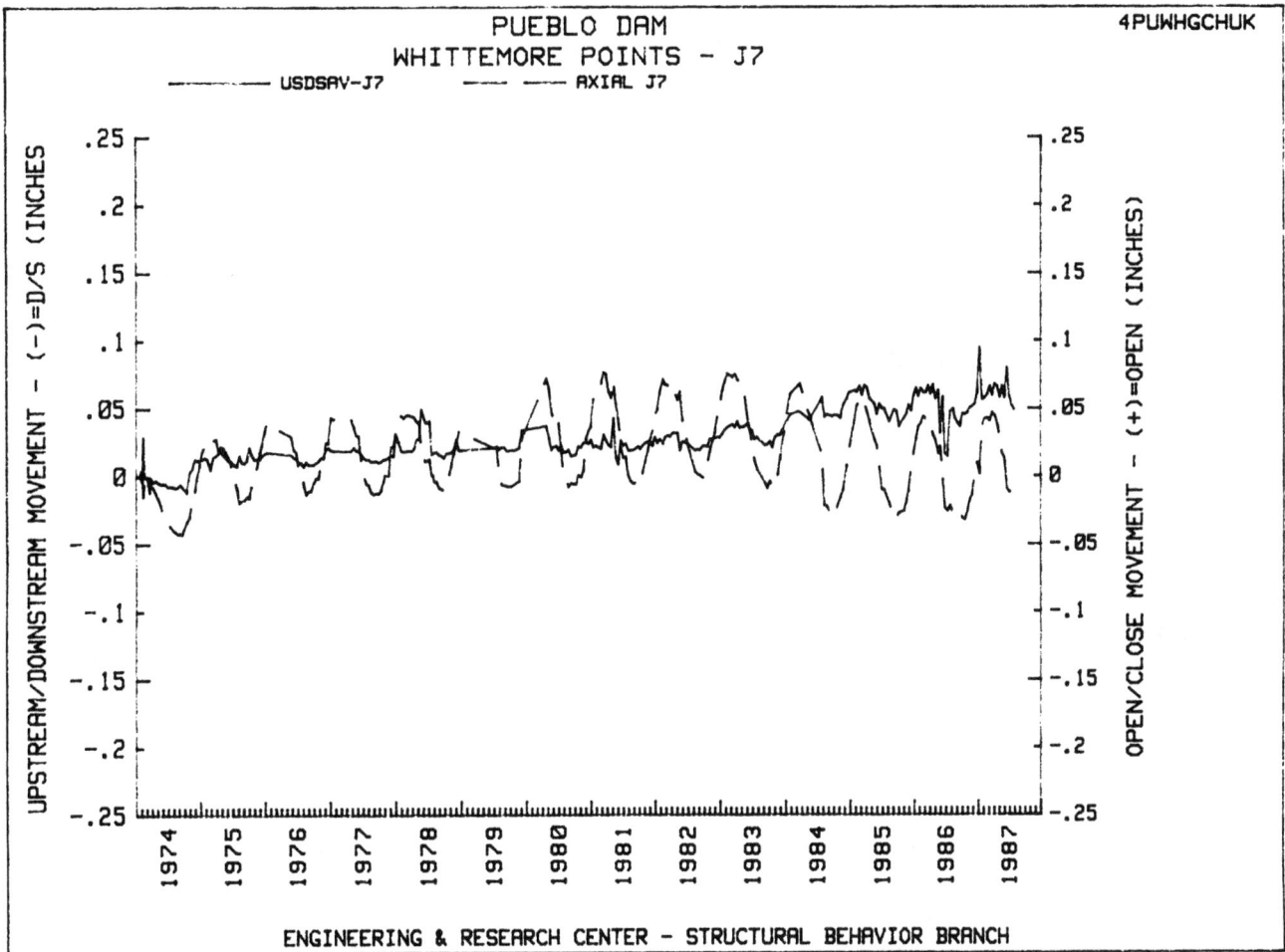

Figure 5-22.—Data plots on crack or joint monitoring.

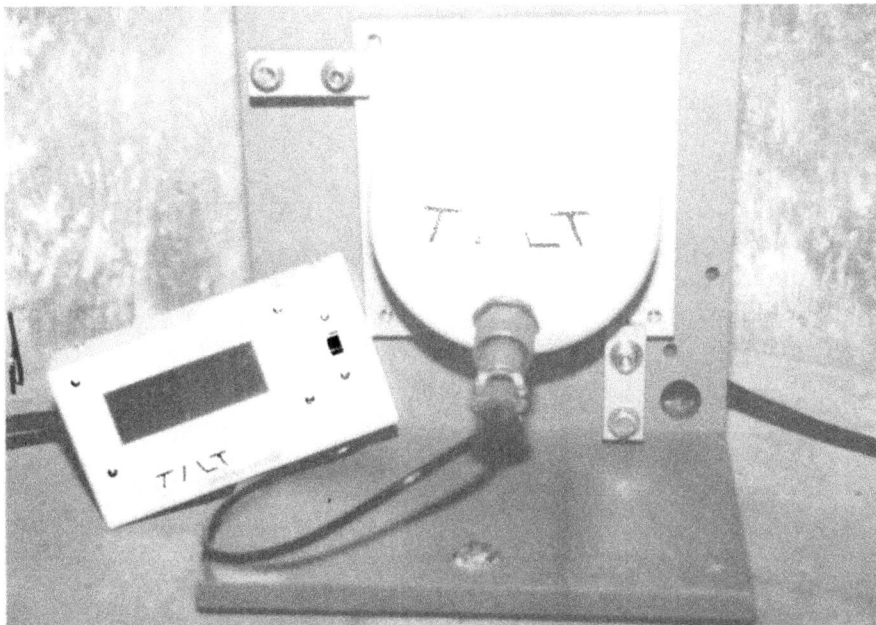

Figure 5-23.—Tiltmeter installed in plumbline well blockout. P801-D-81195.

thoroughly cleaned before the plate is affixed. It is extremely important to align one set of the measuring pegs with the direction to be observed. Initial readings with the sensor are taken as soon as the grout or epoxy has set.

    e. *Monitoring Procedures.*—Tiltmeters are monitored by placing the sensor on the pegs in the same orientation as observed at the time of installation and reading the angle directly on the readout indicator in the "A" position. The switch is then changed to the "B" position and the second direction measured. By using periodic measurements, the rate of angular deformation can be estimated. The data are recorded on a field data form as shown on figure 5-24.

    f. *Maintenance.*—The only maintenance required is the periodic cleaning of the contact points on the ceramic plate and sensor unit and checking the batteries in the readout unit. The sensor unit may be calibrated occasionally by using a tilt of known angle.

    g. *Data Processing and Review.*—The field data are transmitted to the Denver Office where processing and review are accomplished as described in section 1.25.

TILTMETER READINGS

Modified _____8-20-85_____ for
Dam _____RIRIE_____   Date of Observation _____7-14-85_____
Project _____MINIDOKA_____   Observer _____J. STEVENSON_____
Reference Dwg._____   Sheet _____1_____ of _____1_____
                     Res. El. _____

TILTMETER NUMBER	A METER +	B METER +	LOCATION OF METER PLATE
1	0.0150	0.0174	OUTLET WORKS TOWER

Figure 5-24.—Field data form for a tiltmeter device.

Chapter 6

VIBRATION MEASUREMENT DEVICES

A. History and Development of Devices

6.1 General.—Vibrations at damsites may be classified into two principal categories: (1) natural vibrations caused by earthquakes or tectonic movements, and (2) induced vibrations created by operation, maintenance, or construction activities. The effects of both of these categories of vibration are identical, but the magnitude and frequency of the vibrational force may differ a great deal. The possibility of damage such as cracking of the structure or liquefaction of the dam foundation represents a serious condition and an obvious threat to the stabiliy of the dam. Presently, there are no means of controlling earthquake occurrences or of making accurate predictions concerning their occurrence. However, there are positive developments in vibration theory and analysis that are becoming commonly accepted as ways to provide controls on construction blasting. Vibration theory and analysis can aid in the determination of the acceptable amount of explosives, appropriate number and timing of delays, and other essential blasting considerations. The measurement and analysis of construction blasting vibrations are not a part of this manual.

The intensity and duration of ground vibrations that a structure can tolerate without experiencing damage are quite variable. Such physical factors as material type, material density, water content, and natural frequency create the variations that exist in ground vibrations. Not only does physical property variability create problems, but also the parameters that best describe the necessary intensity of ground motion to create structural damage. These parameters include the maximum displacement, maximum particle velocity, maximum acceleration, and the natural frequency, and all may be difficult to determine. The maximum particle velocity is presently the parameter used most extensively as a correlation tool with potential structural damage and correlation charts are available for anticipated damage levels related to particle velocity. However, it should be noted that the particle velocity criteria were developed for structures other than dams and a great deal of care is required to extend this application to dams. The following section applies to all Bureau dams of both concrete and embankment type construction.

6.2 Types of Devices.—Seismic instruments were first developed in the 1930's, while most technological advances have occurred in the past 10 to 15 years. Over the span of 50 years, a number of vibration measuring devices have been developed and, of these, the seismograph (accelerograph) is now the most commonly used seismic instrument. This instrument consists of a sensor (seismometer or accelerometer) and a mechanism for producing a permanent record of the vibration applied to the sensor. Nearly all seismic instruments in use today utilize servo-accelerometers that have the ability to measure motion in a single horizontal, vertical, or transverse plane. Most of the devices used are considered to be "strong-motion" instruments that record significant movement, as opposed to microseismic activity that requires the use of signal conditioners or enhancers to magnify the motion to a recordable level.

The first strong-motion instrument installed by the Bureau in a concrete dam was at Hoover Dam in 1937. Since that time, the additional 9 concrete dams, including Grand Coulee, Hungry Horse, and Flaming Gorge Dams, and 15 embankment dams have been instrumented. Instrumentation at other Bureau dams is in the planning stage.

To develop the methodology, define loading conditions, and verify the results of dynamic analyses, much data are needed on the nature of strong ground motion resulting from earthquakes and the performance of structures (especially dams) subject to earthquake loading. Such data and analyses can be used in the design of new dams and in the modifications of existing dams to provide them with the necessary stability to withstand earthquake loadings appropriate to their locations. In the early 1970's, the Bureau initiated a program to replace the older existing instruments in Bureau structures with new strong motion seismographs and to add instrumentation to structures that do not have it. By 1975, all of the early U.S. Coast and Geodetic Survey type instruments installed on Bureau structures had been replaced by modern, self-contained, three-component, mechanical-optical seismographs. These new instruments are more reliable, and recent modifications and innovations in time coding, power supplies, and installation procedures (borehole as well as surface shelters) allow measurement of response to earthquake strong motions in almost any climatic environment at any location within a structure.

B. Currently Used Devices

6.3 Strong-Motion Earthquake Instruments.—a. *General.*—Strong-motion earthquake instruments, generically termed "vibration monitoring devices," are now being installed at selected Bureau dams and other water-resource related structures. Discussions of site selection factors for these instruments and their distribution in the United States have been made by Viksne [5] and Hudson [6]. The present goal of the Bureau strong-motion program is to continue deploying instruments considering the following factors:

- Seismic zoning and proximity to faults capable of causing earthquakes
- Dimensions of dam and reservoir
- Foundation materials
- Method of construction and type of dam
- New versus existing dams
- Water-resource lifeline features or special interest

The most important factor controlling the Bureau instrumentation program is the location within prioritized seismic zones. The Bureau uses the Uniform Building Code Seismic Zone Map [7] as a basic guideline for this purpose. There is no doubt that such zoning is very broad and that there are "zones within zones" that must be considered in the sequential deployment of strong-motion instruments. Within zones 3 and 4, the zones of highest instrumentation priority, dam locations may vary greatly in respect to proximity of faults capable of causing earthquakes and earthquake magnitude and recurrence exposure levels. For example, within seismic zone 4 in California, the Geological Survey, California Division of Mines and Geology, and the University of Southern California have located arrays of strong-motion instruments where historical seismicity indicates strong-motion events may occur in the relatively near future. An example of a subzone of seismic zone 3, where the Bureau considers significant seismicity possible during the lifetime of its dams, is in east-central California where Boca, Stampede, and Prosser Creek Dams (all embankment dams) are located. Both Boca and Stampede Dams have been instrumented for several years, and the instrumentation of nearby Prosser Creek Dam is now being planned.

The basic zoning program calls for the installation of strong-motion instruments on all existing storage dams in seismic zone 4 and on most of the significant storage dams in zone 3. Newer dams, outside of seismic zones 3 and 4, have been or are planned for strong-motion instrumentation because of engineering concerns for a complete suite of dam instrumentation, including strong motion. This concern may occasionally transcend seismic zone boundary and seismic exposure considerations. However, Bureau instrumentation planning takes into account site-specific detailed seismotectonic studies throughout the 17 Western States. These studies are extending the knowledge of earthquake causative geologic structures such that existing seismic zoning, in terms of magnitude and seismic recurrence to be expected, will require modifications and refinements. Therefore, the Bureau views seismic zone maps as state-of-the-art guidelines that can be modified by engineering judgment. Strong-motion instrumentation, as related to water resource structures throughout the 17 Western States, is concentrated in seismic zones 3 and 4 and widely distributed throughout zone 2.

Present Bureau practice is to consider seismic zoning as the predominant siting factor; however, size, foundation materials, and method of construction are other parameters that may influence strong-motion instrument siting priorities. Foundations that contain unconsolidated silts or fine-grained sands, which make them potentially liquefiable, are also important considerations. Hydraulic fill structures are vulnerable to failures under strong ground motion and may be prioritized for instrumentation over earth embankments placed with modern compaction methods.

b. *Usage of Devices.*—The devices may be categorized according to placement location. Free-field instruments are placed in the vicinity of the dam, but not near enough to be affected by the dam or reservoir; input motion instruments are placed at the downstream toe and abutments to determine the vibration being introduced into the dam; and response instruments are located on the dam to determine dam response to the vibration.

(1) *Free-Field instruments.*—Several sites that the Bureau has "instrumented" have only free-field installations. It is recommended that free-field instruments be installed near both abutments and at the toe

of an "ideal" dam; such instrumentation to be placed at a distance beyond any significant influence of the dam on the recorded ground motion. For structural analyses purposes, the Bureau does not require these optimal free-field installations, nor does the Bureau deliberately attempt to deploy free-field instrumentation, relying instead on other agency network installations. Some Bureau installations are free-field simply because of access or siting limitations around the dams, especially downstream. There are many dams in the United States with near by "free-field" instruments; for example, in California. More study is required to determine the extent of the influence of large structures, reservoirs, and ground water on ground motion.

(2) *Input Motion Instruments.*—The Bureau locates the input motion components of the strong-motion array at the downstream toe and on the abutments as close to the dam as possible. Most of these instruments are placed in prefabricated housing on concrete pads firmly secured to the underlying rock or surficial material. The field review of a few typical dams selected at random will demonstrate that exterior siting conditions are difficult at the toes of most dams due to backwater from downstream reservoirs, tailwater from power-plants, spillway basins, plunge pools, outlet works, or other features not amenable for instrumentation sites. Similar conditions can exist at the dam/abutment contact area where topographic and access considerations frequently allow little room for satisfactory instrument sites. In the abutment areas, an ideal installation in a spatially restricted area would be a small chamber in the natural material where maintenance problems could be minimized. Interior or subsurface input motion sitings consist of borehole instrumentation in the foundation of earth dams and installations in selected galleries of concrete dams. Drainage and grouting galleries, when excavated in earth dam foundations, can be utilized as input motion sites for strong-motion instruments.

Some Bureau embankment dams are built on deep deposits of unconsolidated surficial materials in the bottoms of valleys and canyons. If such deposits represent the typical foundation conditions at the base of a dam, then care is taken to locate the input motion instrument on the surficial material and not, for example, on any bedrock that might be outcropping in the area.

(3) *Response Instrumentation.*—The Bureau ideally installs one or two response instruments on the crest of both earth and concrete dams. The primary location is where maximum deformation during strong motions is expected, usually at the maximum section. A secondary location may be about one-third of the crest length from an abutment, such a location is basically for backup purposes. Many Bureau dams are assymetrical because of differences in abutment slopes, stream channels not in the center of the valley, or other topographical features often related to geological structures. Therefore, the maximum section may be far from the actual center of the dam crest.

If a dynamic analysis of a structure has been done prior to strong-motion instrument deployment, the response instrument locations may be specified based on the analysis. Specified areas would be where lower safety factors and higher loads are expected. These locations would be site specific for each structure. The locations for earth dams depend upon zoning geometry of the dam, types of materials used in the zones, and nature of the foundation; while locations for concrete dams depend upon type of dam (gravity or arch), geometric configuration of the dam, and nature of the foundation materials.

c. *Description of Devices.*—Vibration measuring devices consist basically of a sensor, signal conditioner, and a recorder or storage medium.

(1) *Sensor.*—The sensing devices used are electromechanical units that respond to motion and produce an electrical signal that is, within limits, proportional to displacement, velocity, or acceleration. Most sensors are models of single degree of freedom, spring-mass dashpot systems. The measurement is usually made of the spring extension or compression with the resonant frequency and the damping of the system so proportioned as to produce an electrical signal that is an analog of either acceleration, velocity, or displacement.

Examples of this type of system include accelerometers and velocity gauges, where the spring system is an elastic member with electrical strain gauges attached, and the damping is provided by oil immersion or eddy-current systems; piezoelectric accelerometers, where the spring is an elastic piece of piezoelectric material and the unit is almost undamped; and servo-accelerometers, where the effective spring is a separate electronic system that generates a restoring force similar to that of a spring.

Many other sensor types are available that differ in detail from those previously described. Variable inductance, capacitance, and resistance; or nearly any electrical parameter may be used for the conversion of motion to an electrical analog. However, gauges that do not use the relative displacements of a spring-mass-dashpot system require a fixed point of reference. This requirement severely limits application of this type of gauge.

123

The limits of operation of the single degree of freedom model gauges are usually those of minimum detectable amplitude (sensitivity), maximum amplitude (loss of linear response or mechanical damage), and usable frequence range. It should be noted that some sensors may be described as usable between two given frequency limits with no accompanying warning to the user that the sensor has a highly resonant response above or below those limits. This resonant response must be suppressed by a signal conditioning unit.

To determine the total vector quantity of motion at a certain point, three orthogonal sensing units are usually included in a sensor unit. The magnitude (vector summation) of the motion is the square root of the sum of the squares of the three orthogonal components of motion.

Sensors may be located at the recording unit or placed in a remote location, such as in a drill hole, or elsewhere. Signals are transmitted by a coaxial cable to the signal conditioner. It is common for sensors to contain a starter or triggering device that activates the sensor at some predetermined acceleration, such as 0.01 gravity. The sensor would then continue to operate as long as the motion was greater than that value, and for a short time thereafter.

(2) *Signal Conditioner.*—The term "signal conditioner" refers to all units and devices placed between the sensor and the final output data recorder. These devices are usually power amplifiers that are required to change the micropower signal levels from the sensor to the macrolevels required to activate the recorder system. Signal conditioners also usually include sensitivity controls to permit a desired level of recorded signal.

Signal conditioning equipment may also include analog intergration units for conversion of one measured parameter to another. The equipment may physically be part of the sensor unit, part of the recorder unit, or may be separately packaged.

(3) *Recorder.*—The recorder unit is the final device in the system and is located in a protected environment in a secured area. The recorder presents the output in some usable form for evaluation and/or further use. Most recorders now being used produce a historical record of the input phenomena versus time as a paper record. Such records provide a quick method for visual inspection, and permit a rapid evaluation of peak amplitudes and other values. Detailed study of data in this form requires laborious point-by-point transcription of values for future computation.

Automatic data handling can be obtained for the use of a magnetic tape recorder and later playback onto computing systems. The most desirable system includes an output of both direct reading paper records for rapid field inspection plus a magnetic tape for direct storage, which allows for later computer processing. Most recorders also provide a timing base that is recorded along with the sensor signals as a reference for determining the frequency of the vibration.

d. Operation and Maintenance Considerations.—Optical-mechanical, self-contained accelerographs that record on 70-millimeter photographic film are the most widely used strong-motion devices because of their low cost, high reliability, simple operation, and straightforward maintenance and repair procedures. They are preferred over both analog or digital tape and solid-state recording methods; however, improvements in these alternate types continue to be made. After triggering, up to 25 cumulative minutes of motions equal or stronger than 0.01 gravity can be stored. In most units, recording continues for 10 seconds after motions drop below the 0.1-gravity level.

Several calibration tests, both laboratory and field, must be conducted before the instruments are installed. Instrument sensitivity, or the deflection of the seismograph trace under known loading, is generally determined by the manufacturer under laboratory conditions before the equipment is brought to the field. Calibration tests for damping and natural frequency are easily performed in the field. These calibration tests are recorded at the beginning and end of each roll of film.

Two timing traces are continuously recorded with earthquake ground motions, with one trace always being a square wave pulse produced internally. The second trace may also be a square wave pulse, a binary time code controlled by an oscillator and set according to an external clock or, more commonly, a WWVB Bureau of Standards radio signal time code. A timing trace is necessary so that each motion even can be clearly identified for correlation with events on other records.

Most of the maintenance of Bureau strong-motion instruments is accomplished through a cooperative effort between the Bureau and the USGS (United States Geological Survey). Typically, a technician checks the instrument installation after a known strong-motion event, or every 6 months, and replaces the film. If a strong-motion event has been recorded, the USGS routinely digitizes the analog record; corrects for film distortion, variable film speed, and instrument response; and filters the record to remove long-period noise.

The maintenance of the Bureau's strong-motion arrays is complicated by the diversity of the sites, and with extremes such as high elevation and very cold to very hot climates. The addition of sunroofs and improved air circulation by venting has lessened the heat problems, while electric heating and insulated shelters have dealt with extremely cold conditions. Drainage is also an important aspect of surface instrument sites as poor drainage has occasionally caused temporary operational problems and shutdowns. In attempts to located instrument housings out of the way of access roads and work areas, the housings have been located close to cut slopes; however, sloughing material and minor slides have interfered with the operations. Small concrete revetments are now being used to protect the housing.

    e. *Monitoring Procedures.*—The monitoring procedure consists of securing the paper records of the tracings from the data recorder on a periodic basis. Additional data are secured from the recorder immediately following known seismic events. Monitoring of seismic devices on Bureau projects is usually done by USGS personnel.

    f. *Data Processing and Review.*—The data are analyzed by USGS personnel in conjunction with Bureau geological personnel. If the data indicate a problem at a damsite, procedures for review and necessary action as described in section 1.25 should be followed.

Chapter 7

SUPPLEMENTAL MEASUREMENTS AND DEVICES

A. Temperature Measurements

7.1 General.—A temperature rise occurs in concrete after placement due to heat of hydration of the cement. To aid in controlling this temperature rise, many concrete dams use embedded cooling pipes through which cool water is circulated during and after construction.

After the peak temperature is reached, the internal temperature will decline with time depending upon thickness of the section, exposure conditions, rate and amount of continued heat of hydration, and whether artificial cooling is continued. The peak temperature is usually reached between 7 and 20 days in massive concrete sections where no artificial cooling is used. These sections may maintain their maximum temperature for several weeks, after which the temperature will drop slowly over several years. In thin structures or structures using artificial cooling, the peak temperature is usually reached in 2 to 6 days after placement. Temperature change is ideally limited to about 1 °F or 1°C per day; however, in thin structures exposed to very cold temperatures, it may approach 4 °F or 4 °C per day.

Timely operation of the embedded cooling system will reduce the tendency of the concrete to crack during the construction period. Similarly, cooling should not proceed too rapidly or shrinkage cracking may result. In some instances, the exterior of the dam is insulated while the interior is cooled gradually. Regardless of the method of cooling used, it is important that the dam be cooled prior to grouting of the contraction joints.

After a dam is in operation, warm sunlight on the downstream face can create strains that may result in differential stresses between the downstream face and the cooler upstream face that is mostly below water. Conversely, during the winter, the relatively warm water on the upstream face may cause differential stresses versus the downstream face that may be exposed to very cold air temperatures.

As previously discussed, many of the stress and strain measuring devices are also capable of measuring temperature; however, there are several devices that measure only temperature. These devices are embedded in the internal portions and near the surfaces of dams. They monitor the temperature so that temperature-induced movements and loadings can be correlated and corrections to other instrumentation can be made.

7.2 Resistance Thermometers.—**a.** *Usage.*—Resistance thermometers have commonly been used for long-term temperature measurements in Bureau concrete dams, while thermocouple devices have been used for short-term, construction period observations. Thermocouple devices are discussed in section 7.3.

Temperature-sensing devices are usually installed on a grid pattern in the maximum section or in several sections of a dam to measure distribution of temperature. Such measurements are of great importance because the volume change caused by temperature fluctuations is one of the factors that contributes significantly to stress and deflection. These devices are also used during construction to facilitate control of the cooling system operations. Another extensive use of these devices has been the development of concrete temperature histories to study the heat of hydration, and to evaluate conditions that contribute to or accompany the formation of thermal cracking in mass concrete.

Concrete surface temperature-sensing devices are embedded at various locations on the downstream face and at selected vertical intervals between the base and crest on the upstream face. The latter installations furnish information on temperature variations in the reservoir. Figures 7-1 and 7-2 show typical thermometer installations in gravity and arch dams, respectively.

b. *Description of Devices.*—The resistance thermometer operates on the principal of changing resistance in a copper wire as the temperature changes. The thermometer device shown on figure 7-3 is read with a Wheatstone bridge-type readout box.

c. *Installation Procedures.*—(1) *Preliminary.*—After receipt of shipment, the thermometers should be checked for damage, and initial readings should be taken with the thermometer submerged in a bucket of water at constant temperature.

If cable splicing is necessary in the field, calibration should be accomplished in a manner similar to that for strain meters (sec. 4.7 d). This is done by immersing the units in a bucket of water and taking readings both before and after cable splicing. The corrected resistance calibration of the meter at the basic temperature,

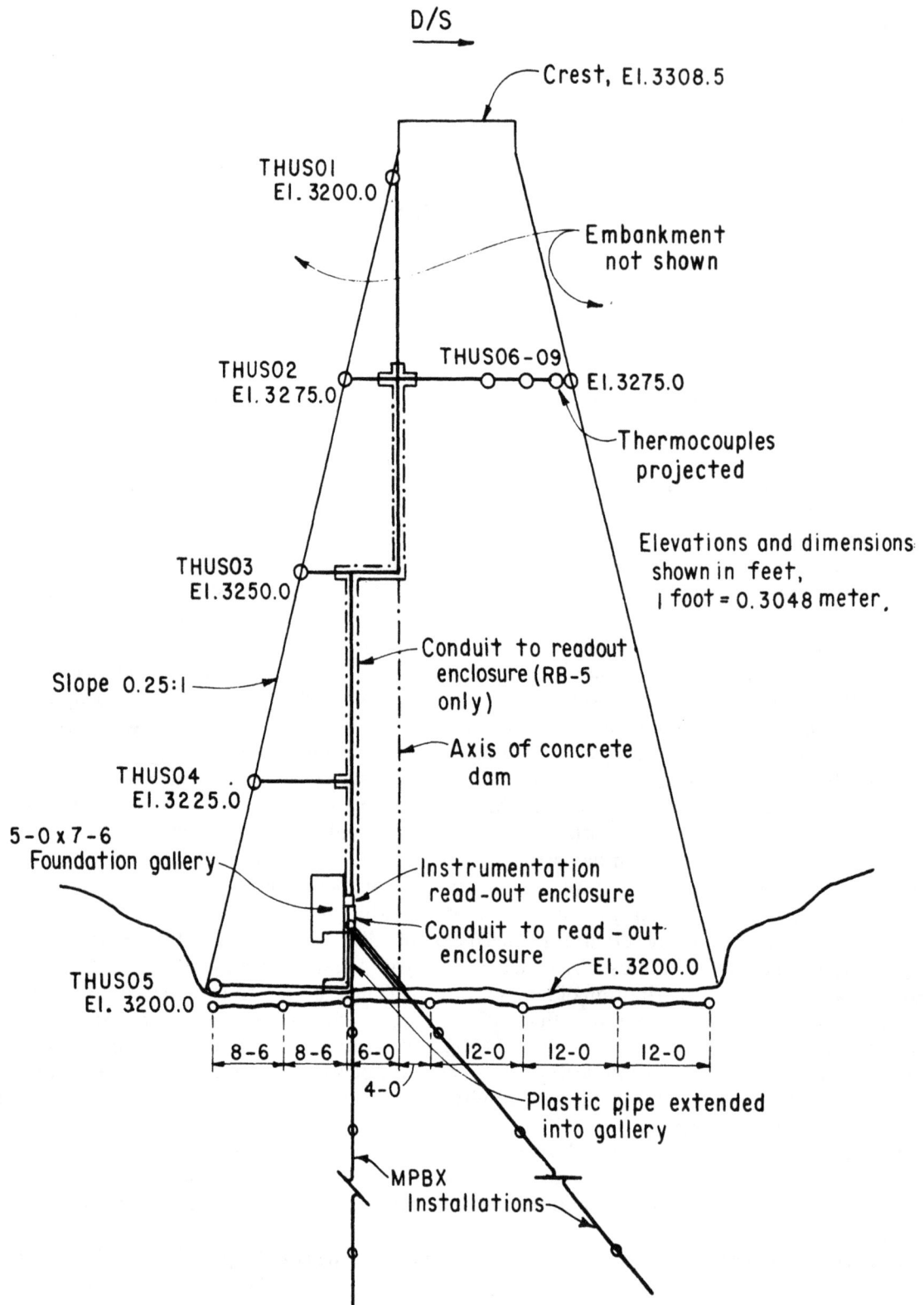

Figure 7-1.—Typical thermometer installation on a gravity dam. From 963-D-121.

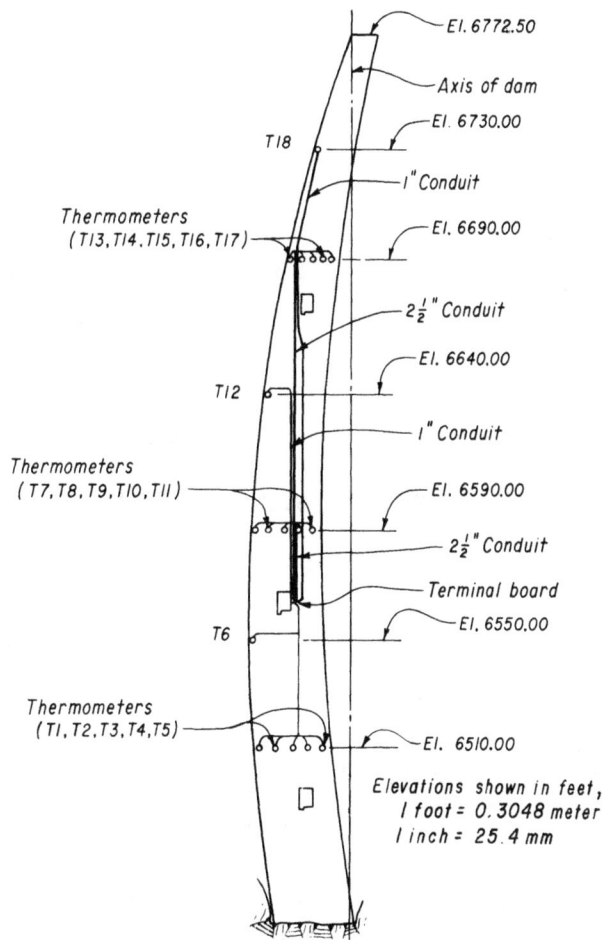

Figure 7-2.—Typical thermometer installation on an arch dam. 622-D-2807.

usually in ohms at 0 °F (−18 °C), is obtain by adding algebraically the quantity ($R' − R$) to the basic resistance value furnished by the manufacturer, where:

R' = resistance reading at end of spliced cable, and
R = resistance reading at splice.

A final adjustment of the temperature factor will be determined after installation when the cable leads are trimmed to the proper length. The previously corrected temperature factor is then adjusted by algebracially adding to the factor the measured instrument resistance, after trimming, minus the measured instrument resistance before trimming.

If a three-wire system is used, the resistance of the lead wire is compensated for by a readout box Wheatstone bridge and, therefore, no correction is necessary.

(2) *Installation.*—Embedment of a resistance thermometer is a relatively simple procedure because orientation is usually not important, and care in concrete placement around the device is less critical than with the installation of strain or joint meters. When location of the instrument relative to the top or bottom of a lift is not important, the thermometer is simply laid in a shallow hole 6 to 8 inches (152 to 203 mm) deep, covered immediately with fresh concrete, and the area lightly vibrated. Installation of several or a group of thermometers on a single horizontal plane within a lift is usually accomplished by placing them at the bottom of the lift. The thermometers are taped or tied securely with wires embedded in the top of the previous lift at the approximate locations desired.

For accurate spacing of thermometers at various heights in a lift, the meters may be fastened to a pole or rod that has been embedded in the previous lift. Either wood or plastic poles are preferred because of their favorable thermal properties; reinforcing bars should not be used for this purpose.

129

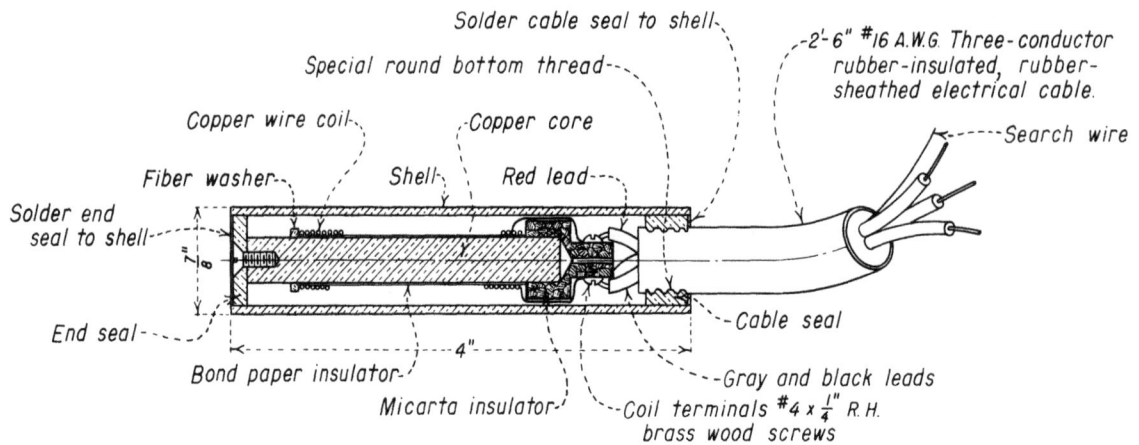

EMBEDDED TYPE RESISTANCE THERMOMETER

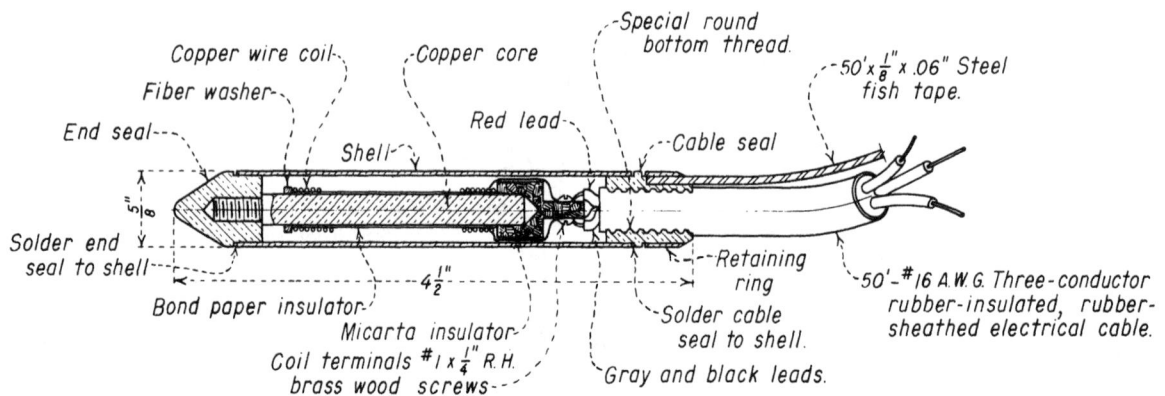

INSERTED TYPE RESISTANCE THERMOMETER

NOTES

Resistance coils are #33 A.W.G. enameled copper wire,
    noninductively wound, 46.00 ohms at 70.0°F.
End seal, cable seal, and shell are brass.
Space between thermometer coil and shell is filled
    with G.E.-227 sealing compound prior to soldering
    end seal.
For insert thermometer: Tee-shaped end of fish tape
    is fitted to tee-slot in cable seal. Retaining ring
    pressed over cable seal holds fish tape in place.

Figure 7-3.—Details of a resistance thermometer. 40-D-5088.

Thermometers located within about 3 feet (0.9 m) of exposed concrete surfaces or bulkhead faces that will be subject to daily temperature variations must be placed accurately and tied securely at their intended distance from the surface. It is most desirable to use ties or frames that have a similar thermal conductivity as the concrete whenever possible.

Generally, thermometers are not located near strain or other meters because the temperature at those locations can be measured by those devices.

d. Monitoring Procedures.—The thermometers are read at desired intervals using a standard Wheatstone bridge-type readout device as described in section 4.7 e. The data are recorded on a field data form as shown on figure 7-4.

e. Maintenance.—No maintenance other than that for the readout unit described in section 4.7 f is possible.

130

UNITED STATES
DEPARTMENT OF THE INTERIOR
BUREAU OF RECLAMATION
COLUMBIA BASIN PROJECT
GRAND COULEE THIRD POWER PLANT
FOREBAY DAM
METER NOTES

Observer A.W. Recorder J.C. Time 0800 To 1100 Date 5-27-87

THERMO.	3-L. RES	TEMP. DEG. F.	METER	TOTAL RESISTANCE	RESISTANCE RATIO	REVERSE RATIO	
T 1	44.73	57.3	J 1				
T 2	44.79	57.9	J 2				
T 3	44.83	58.3	J 3				
T 4	44.88	58.8	J 4				
T 5	44.45	54.5	J 5				
T 6	44.41	54.1	J 6				
T 7	44.48	54.8	J 7				
T 8	44.75	57.5	J 8				
T 9	45.40	64.0					
T 10	45.96	69.6					
T 11	45.99	69.9					
T 12	45.31	63.1					
T 13	45.20	62.0					
T 14	45.42	64.2					
T 15	45.50	65.0					

Remarks: _____

STD. COIL	BEGIN	END	TEST SET SERIAL NO. 44914
RES. RATIO	1.0303	—	
REV. RATIO	0.9707	—	

Figure 7-4.—Field data form for thermometers. 1222-D-2443.

131

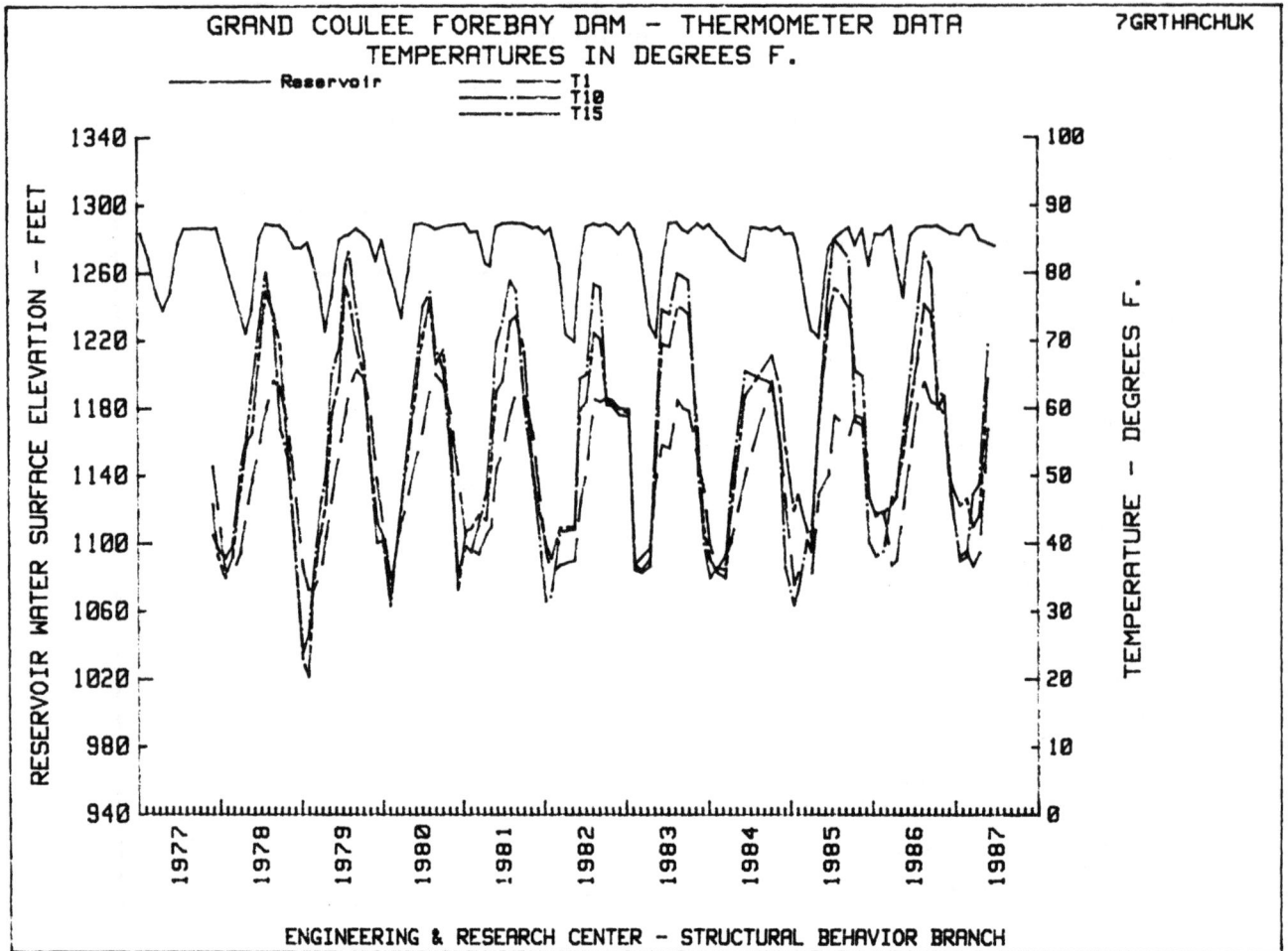

GRAND COULEE FOREBAY DAM – THERMOMETER DATA
TEMPERATURES IN DEGREES F.

Figure 7-5.—Data plots on thermometers.

f. Data Processing and Review.—The field data forms are transmitted to the Denver Office for processing and review as discussed in section 1.25. Data plots are prepared as shown on figure 7-5.

7.3 Thermocouples.—a. Usage.—Thermocouples are suitable for measurement of temperatures in massive concrete structures under certain conditions. Uses include detailed temperature histories over the first few hours or days after placement, temperature variations near bulkhead and horizontal lift surfaces during the exposure interval, temperature variations adjacent to forms and formed surfaces, and daily temperature cycles close to exterior faces of the structure.

The accuracy of the temperature values determined within a given temperature range depends primarily upon the sensitivity and proper operation of the indicating instrument. For the usual temperatures encountered in massive concrete construction, which is 0 to 150 °F (−18 to 66 °C), accuracies may range from about 0.1 °F (0.05 °C) with precise equipment under favorable circumstances to about 1 °F (0.6 °C) with some commercially available indicating potentiometers under normal operating conditions.

b. Advantages and Limitations.—The thermocouple is considered to be the most versatile of temperature indicators for engineering applications. Different types of thermocouples have been used for measuring temperatures from nearly absolute zero up to 5400 °F (2982 °C). The sensing element is small, which permits measurements at a relatively small point, and has a correspondingly low thermal capacity that contributes toward rapid response to temperature variations.

While the thermocouple is inexpensive, simple to construct, and may be made as rugged as necessary, the indicating instruments required for readout are comparatively elaborate. The thermocouple circuit is subject to parasitic and spurious voltages, its precision is reduced an indeterminate amount by the inherent non-homogeneity in the extension cables, and the extremely small voltage signals generated demand highly sensitive measuring instruments that are also subject to extraneous influences. Electronic indicating potentiometers

132

used for commercial and industrial control applications usually result in decreased precision and an increase in servicing problems. Generally, resistance thermometers are preferred over thermocouples in concrete installations because they have been found to be more dependable, generally longer lasting, of greater precision, and somewhat less complicated in operation.

c. *Description of Devices.*—Thermocouples operate on the principle that when two dissimilar metal wires are joined together, a change in temperature produces a change in voltage at the junction. The magnitude of the voltage depends on the metals used, and is proportional to the temperature at the junction. A suitable potentiometer is used for detecting the voltage and thereby determining the temperature. The material most used for concrete dam temperature monitoring is a copper constantan combination, which is known as a type "T" thermocouple, and has a temperature range from subzero to +700 °F (+371 °C). Other common types of thermocouples occasionally used include the "J" (iron constantan) with a temperature range of +32 to 1400 °F (+0 to 700 °C) and "K" type (chromel and alumel) with a range of −330 to +2300 °F (−201 to 1260 °C).

d. *Installation Procedures.*—Installation of thermocouples is accomplished by the same procedures as those for resistance thermometers discussed in section 7.2.

e. *Monitoring Procedures.*—Monitoring of thermocouples is accomplished using a special readout device that measures temperature directly.

f. *Maintenance.*—No maintenance is possible except for checking the accuracy and batteries in the readout unit. Data are recorded on field data forms as shown on figure 7-6.

g. *Data Processing and Review.*—The data forms are transmitted to the Denver Office for processing and review as discussed in section 1.25. Data plots are prepared as shown on figure 7-7.

B. Special Measurements

7.4 Water Quality Testing.—a. *Purpose.*—Water quality testing is conducted on water samples obtained from many Bureau embankment damsites and from one concrete dam, Elephant Butte Dam in New Mexico. The usual reason for conducting a water quality testing program at a dam is the desire to compare the characteristics of seepage water to the characteristics of the reservoir water. Such comparisons may be valuable in the detection of factors affecting the safety of dams. These factors include:

- Detection of possible solutioning of foundation or abutment rock or chemically bound aggregated soil materials. Solutioning, which is the process whereby water flowing through a material chemically dissolves a portion of that material, can result in a weakening of a dam's foundation or abutments and/or increase the size of fissures or cracks where water can flow, thereby leading to increased seepage quantities.
- Progressive erosion of soil particles from foundations and abutments. If seepage water contains a large or increasing quantity of soil particles, it is an indication that internal erosion is occurring. Such erosion can lead to a form of progressive erosion failure known as piping.
- Detection of new seepage paths. If flow increases in certain drains and the water contains new or increased quantities of certain constituents, an indication exists that a new seepage path may have developed.

b. *Water Sampling Techniques.*—The data obtained from a water testing program is of no significant value if the sampling techniques used for obtaining the water specimens, sample handling techniques, and testing methods are not in accordance with established practices. In fact, erroneous data would probably be more detrimental to proper decision making than no data at all. Therefore, it is essential that all phases of the testing program be properly conducted to eliminate any doubt regarding how representative a sample may be and whether it was properly handled and tested. The Applied Sciences Branch, Division of Research and Laboratory Services, at the Engineering and Research Center in Denver, Colorado is available for consultation and/or field participation to advise on sampling techniques and analysis requests. Field personnel are encouraged to utilize this pool of expertise.

Recommended sampling, handling, and testing procedures are given in two Bureau publications, the Earth Manual [8] and the Concrete Manual [9]. Other sources for this type of information would include the ASTM (American Society for Testing Materials), APHA (American Public Health Association), EPA (Environmental Protection Agency), and the USGS (United States Geological Survey). No attempt is made in this manual to duplicate or quote these recommendations, except to highlight general guidelines as they affect Bureau water testing programs.

```
                    EMBEDDED THERMOCOUPLE TEMPERATUARES
                          03/30/87 - 05/31/87

Block 12-9M (Sta. 93+02 12.6' U/S) THUS03       Block 12-9M (Sta. 93+02 12.6' U/S) THUS03

DAY  DATE  TIME  TEMP                        DAY  DATE  TIME  TEMP
                 (°F)       REMARKS                          (°F)       REMARKS
145  03/30 1730  47.1                        180  05/14 1630  68.2
146  03/31 1630  48.6                        181  05/15 ----  ----
147  04/01 1800  50.0                        182  05/16 ----  ----
148  04/02 1700  52.0                        183  05/17 ----  ----
149  04/03 1500  51.0                        184  05/18 ----  ----
150  04/04 ----  ----                        185  05/19 ----  ----
151  04/05 ----  ----                        186  05/20 1530  68.3
152  04/06 1630  51.6                        187  05/21 0930  68.8
153  04/07 1630  50.6                        188  05/22 0900  70.3
154  04/08 1430  50.2                        189  05/23 ----  ----
155  04/09 1600  52.2                        190  05/24 ----  ----
156  04/10 1430  54.7                        191  05/25 ----  ----
157  04/11 ----  ----                        192  05/26 1500  69.0
158  04/12 ----  ----                        193  05/27 1300  70.6
159  04/13 1630  57.1                        194  05/28 1400  72.0
160  04/14 1230  57.9                        195  05/29 1330  72.0
161  04/15 1300  58.4                        196  05/30 1000  73.0
162  04/16 1300  60.4                        197  05/31 ----  ----
163  04/17 1530  60.1
164  04/18 ----  ----
165  04/19 ----  ----
166  04/20 1300  65.3
167  04/21 1330  65.3
168  04/22 1300  61.8
169  04/23 1350  62.0
170  04/24 1600  63.5
171  04/25 ----  ----
172  04/26 ----  ----
173  04/27 ----  ----
174  04/28 0830  63.4
175  04/29 1400  62.8
176  04/30 1400  63.0
177  05/01 0930  63.5
178  05/02 ----  ----
179  05/03 ----  ----
180  05/04 1430  64.9
181  05/05 1000  65.3
181  05/06 1430  65.9
181  05/07 1530  64.9
182  05/08 0900  65.8
183  05/09 ----  ----
184  05/10 ----  ----
185  05/11 1300  68.4
186  05/12 1600  68.4
187  05/13 1500  69.3

Note:  For previous readings see Weekly dated January 7 - January 13, 1987
```

Figure 7-6.—Field data form for thermocouples.

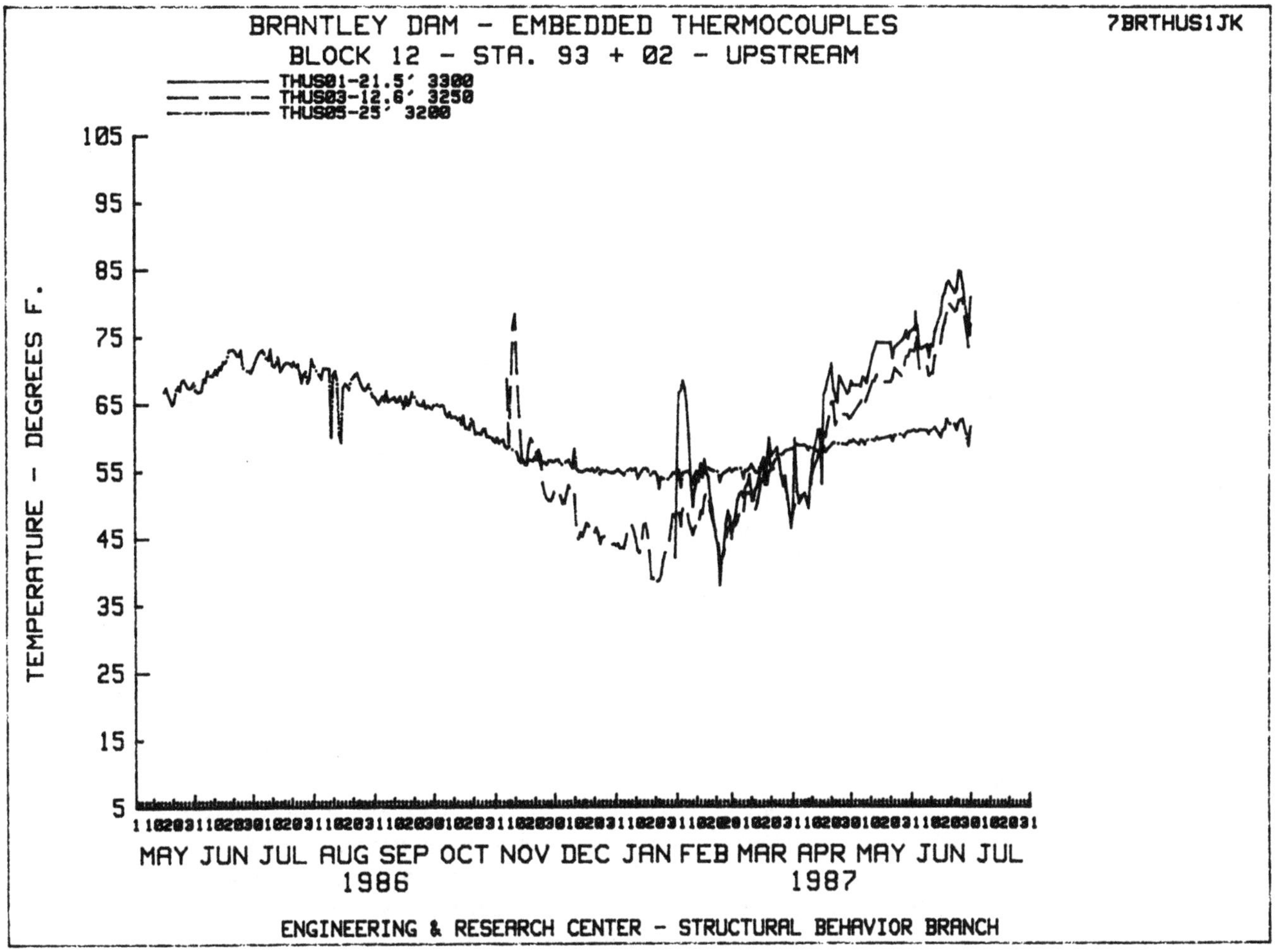

BRANTLEY DAM - EMBEDDED THERMOCOUPLES
BLOCK 12 - STA. 93 + 02 - UPSTREAM

Figure 7-7.—Data plots on thermocouples.

(1) *Sample Containers.*—Glass or plastic sample containers ranging in size from 1 pint to ½ gallon (0.5 to 1.9 L) may be supplied by the chemistry laboratory. These containers will have been thoroughly cleaned, sterilized, and sealed prior to shipment to the site. A small piece of tape placed over the container cap will indicate whether the container has been opened (and possibly contaminated) prior to the actual sampling time. A container that has been previously opened or is not properly sealed should not be used. Field personnel should not attempt to clean or sterilize any sample container for later use in sampling.

(2) *Shipping Containers.*—Sample bottles are shipped or delivered to the damsite in special insulated coolers designed to protect the sample from breakage and to offer protection from freezing or excessive heat. It is intended that the same or similar shipping container be used for returning the full sample bottles. Regardless of the degree of protection offered by the shipping container against damage, it is important to use all possible care while handling and transporting the samples to minimize the possibility of such damage. Bureau samples should be shipped directly to the Bureau of Reclamation, Applied Sciences Branch, at the Engineering and Research Center in Denver, or to the contract laboratory.

(3) *Sampling.*—The purpose of sampling water at a dam is to enable determination of the exact characteristics (physical and chemical) of the water at the location sampled under the existing conditions (temperature, flow, etc.) at the time of sampling. Therefore, the sample must be truly representative of the water existing at that time and place and, to accomplish this, standard sampling methods must be used. Sampling from observation wells or vertical pipes may be accomplished:

(a) Determine existing water level in pipe and record on field data form.
(b) Using either a clean baler or pump, lower water level in pipe to bottom of pipe or by at least 10 feet (3.05 m), unless rapid inflow prevents doing so.

(c) Allow sufficient time for water level to return to at least one-half of the amount of depth removed.

(d) Lower a clean sampling device to level desired and secure a water sample. Many commercial water sampling devices are available, and most of them provide for a plug or stopper on the sampler that can be removed by pulling on a string after the weighted sampler has been lowered to desired depth. To avoid sampling any floating debris, oil, etc., on the water surface, sampling at the water-level surface should not be done.

Sampling from protruding drainpipes or from free-flow weirs should be accomplished by sampling directly into a hand-held sample container or by using an extension arm that is hand controlled. The container should be drawn across the entire width of flow with a quick sideway motion that is repeated until the container is filled. It is important to determine the flow rate at the time of sampling so that the flow data can be entered on the field data form.

Sampling from open channels, ditches, or open drains should be preceded by a determination of the flow rate. For a small channel or low flow, sampling is then accomplished directly into a hand-held container. A mechanical extension arm may be used to hold the container if access by hand is difficult. The sample container should be lowered by hand into the water to some point below the water surface while using the other hand to hold the lid on the container. At the desired depth, the lid is removed and the container allowed to fill. At all times, the mouth of the container should be oriented downstream and the bottom of the container upstream. Sampling on the bottom of a channel may allow some bottom sediment to enter the sample, and a surface sample may allow floating debris to contaminate the sample; avoid sampling these areas.

In large or deep channels or when sampling from the reservoir, sampling is done using the same type of clean sampling device previously discussed. The device is lowered to the depth desired, stopper pulled, device allowed to fill, and then the device is removed from the water body. The device is then emptied into a clean sample container. This operation is repeated until the sample container is full. Each sample bottle should be labeled with the dam name, date, and sample location. Measurements of flow, depth of sampling, date, and sampling location should be recorded on the field data form (fig. 7-8).

Samples may be temporarily stored in clean, secure areas at temperatures that are not allowed to fall below freezing or to rise high enough so that the chemistry of the sample may be altered. The EPA methods manual should be consulted for maximum storage times and allowable temperature ranges for both temporary and long term sample storage. For some tests, it is recommended that certain additional chemicals be added to the sample to fix certain constituents. Sampling instructions for each dam shall state type and amounts of chemicals to be added to the samples.

c. *Type of Tests.*—Both field and laboratory testing is usually performed for each water testing program. While most water quality tests could theoretically be performed in a small field laboratory, it is customary for the majority of the tests to be conducted at the Bureau's Engineering and Research Center in Denver or in a regional laboratory. Naturally, some tests are preferably performed in the field because of their ease of performance or the fear that the test values may change over a period of time. These would include tests for pH, temperature, conductivity or specific conductance and, occasionally, turbidity.

Laboratory tests would include pH, conductivity or specific conductance, total dissolved solids, total suspended solids, total cations plus anions, calcium, magnesium, sodium, potassium, carbonate, bicarbonate, sulfate, and chloride. Specialized tests for special site studies are also sometimes conducted. Laboratory test reports, such as those illustrated on figures 7-9 and 7-10, are then prepared and distributed.

d. *Data Analysis and Presentation.*—Data from the laboratory analysis on seepage water are compared to the laboratory analysis on the reservoir water. From this comparison, it may be determined if water seeping through the foundation or abutments is dissolving any of the material through which it is passing, or if erosion is occuring. If the original data are available, it is desirable to compare the seepage water characteristics with the characteristics of the ground water or river water prior to dam construction. This comparison would help determine what changes in seepage patterns have occurred due to dam construction.

One commonly used method for water quality data presentation is the Stiff diagram (fig. 7-10). These graphical plots allow for the changes in water chemistry to be visually confirmed in a quick and direct manner. Computer programs that plot Stiff diagrams are available on the Bureau's CYBER and Hewlett-Packard computer systems. These diagrams are regularly prepared for illustration of major cation and anion content. Water quality analysis reports are also plotted as shown on figure 7-11.

For more complex data analysis, where a number of sources of water need to be evaluated and compared, it is recommended that multivariate statistical analysis be used on data sets from a well-designed sampling program. Examples of such analyses are cluster analysis, principal component analysis, and discriminant

```
                USBR CHEMISTRY LABORATORY
              ENGINEERING AND RESEARCH CENTER
               PO BOX 25007/MAIL CODE D-1523
                   DENVER, COLORADO 80225
                 FTS-776-6201/303-236-6201
                          10/24/85

--------------------------------------------------------------------------------
******************* WATER QUALITY SAMPLE IDENTIFICATION ********************
--------------------------------------------------------------------------------

Project........................Elephant Butte
Sampling date..................7/11/85
Number of samples.............. 13
Chemistry lab numbers..........E- 8951 to 8963
Analyst........................B.J. Frost
Stored under file name.........SW8951
--------------------------------------------------------------------------------

SAMPLE  1     E-8951     #1 Reservoir water @ L. abutment

SAMPLE  2     E-8952     #2 Dirt Dike #1

SAMPLE  3     E-8953     #3 Dirt Dike #2

SAMPLE  4     E-8954     #4 R. abutment - Lake @ dam

SAMPLE  5     E-8955     #5 L. abutment - Lake @ dam

SAMPLE  6     E-8956     #6 R. abutment Drain #3

SAMPLE  7     E-8957     #7 R. abutment 'X' Drain

SAMPLE  8     E-8958     #8 'D' Gallery F12-104 North

SAMPLE  9     E-8959     #9 'D' Gallery F12-104 South

SAMPLE 10     E-8960     #10 L. abutment Drain #10

SAMPLE 11     E-8961     #11 'A' Gallery 58-66 North

SAMPLE 12     E-8962     #12 'A' Gallery 58-66 South

SAMPLE 13     E-8963     #13 Rio Grande 1000' d/s from dam

--------------------------------------------------------------------------------
```

Figure 7-8.—Field data form for sampling of water.

analysis, all available on the CYBER system. The Applied Sciences Branch at the Engineering and Research Center in Denver will assist all Bureau offices in usage of these programs. It is important to note that no amount of graphics or statistical manipulation can correct a poorly designed sampling program or an improper analysis request.

C. Special Devices

7.5 Automated Devices.—Recent trends in automated instrumentation programs at Bureau dams have been influenced by the reduction of field forces, situations requiring frequent monitoring, need for timely data acquisition and analysis, and the volume of data generated by the instrumentation required at these dams. For these and other reasons, the decision to automate some or all of the instrumentation at a site may be made.

```
                            USBR CHEMISTRY LABORATORY
                      REPORT OF MAJOR CATION AND ANION ANALYSES
                                   10/24/85
--------------------------------------------------------------------------------
PROJECT................................ Elephant Butte
SAMPLING DATE.......................... 7/11/85
--------------------------------------------------------------------------------


E- 8951    #1 Reservoir water @ L. abutment

pH..................................... 8.20E+00
Conductivity........................... 5.08E+02  microsiemens @ 25
Suspended solids....................... 5.52E+00  mg/L
Dissolved solids-105C.................. 3.04E+02  mg/L
Sum of cations+anions.................. 3.55E+02  mg/L


Calcium................. 2.15E+00  meq/L  ......... 4.30E+01  mg/L
Magnesium............... 7.16E-01  meq/L  ......... 8.70E+00  mg/L
Sodium.................. 1.96E+00  meq/L  ......... 4.50E+01  mg/L
Potassium............... 1.08E-01  meq/L  ......... 4.23E+00  mg/L
Carbonate............... 0.00E+00  meq/L  ......... 0.00E+00  mg/L
Bicarbonate............. 2.23E+00  meq/L  ......... 1.36E+02  mg/L
Sulfate................. 2.03E+00  meq/L  ......... 9.75E+01  mg/L
Chloride................ 5.84E-01  meq/L  ......... 2.07E+01  mg/L
--------------------------------------------------------------------------------


E- 8952    #2 Dirt Dike #1

pH..................................... 8.10E+00
Conductivity........................... 1.18E+03  microsiemens @ 25
Suspended solids....................... 3.90E-01  mg/L
Dissolved solids-105C.................. 7.49E+02  mg/L
Sum of cations+anions.................. 8.26E+02  mg/L


Calcium................. 3.53E+00  meq/L  ......... 7.08E+01  mg/L
Magnesium............... 1.18E+00  meq/L  ......... 1.43E+01  mg/L
Sodium.................. 6.22E+00  meq/L  ......... 1.43E+02  mg/L
Potassium............... 4.76E-02  meq/L  ......... 1.86E+00  mg/L
Carbonate............... 0.00E+00  meq/L  ......... 0.00E+00  mg/L
Bicarbonate............. 5.09E+00  meq/L  ......... 3.10E+02  mg/L
Sulfate................. 4.85E+00  meq/L  ......... 2.33E+02  mg/L
Chloride................ 1.49E+00  meq/L  ......... 5.28E+01  mg/L
--------------------------------------------------------------------------------


E- 8953    #3 Dirt Dike #2

pH..................................... 8.30E+00
Conductivity........................... 2.45E+03  microsiemens @ 25
Suspended solids....................... 1.62E+04  mg/L
Dissolved solids-105C.................. 1.72E+03  mg/L
Sum of cations+anions.................. 1.74E+03  mg/L


Calcium................. 4.47E+00  meq/L  ......... 8.95E+01  mg/L
Magnesium............... 3.09E+00  meq/L  ......... 3.76E+01  mg/L
Sodium.................. 1.79E+01  meq/L  ......... 4.11E+02  mg/L
Potassium............... 1.13E-01  meq/L  ......... 4.43E+00  mg/L
Carbonate............... 0.00E+00  meq/L  ......... 0.00E+00  mg/L
Bicarbonate............. 5.31E+00  meq/L  ......... 3.24E+02  mg/L
Sulfate................. 1.52E+01  meq/L  ......... 7.31E+02  mg/L
Chloride................ 4.12E+00  meq/L  ......... 1.46E+02  mg/L
--------------------------------------------------------------------------------
```

Figure 7-9.—Typical water sampling printout on the analyses of major cations and anions.

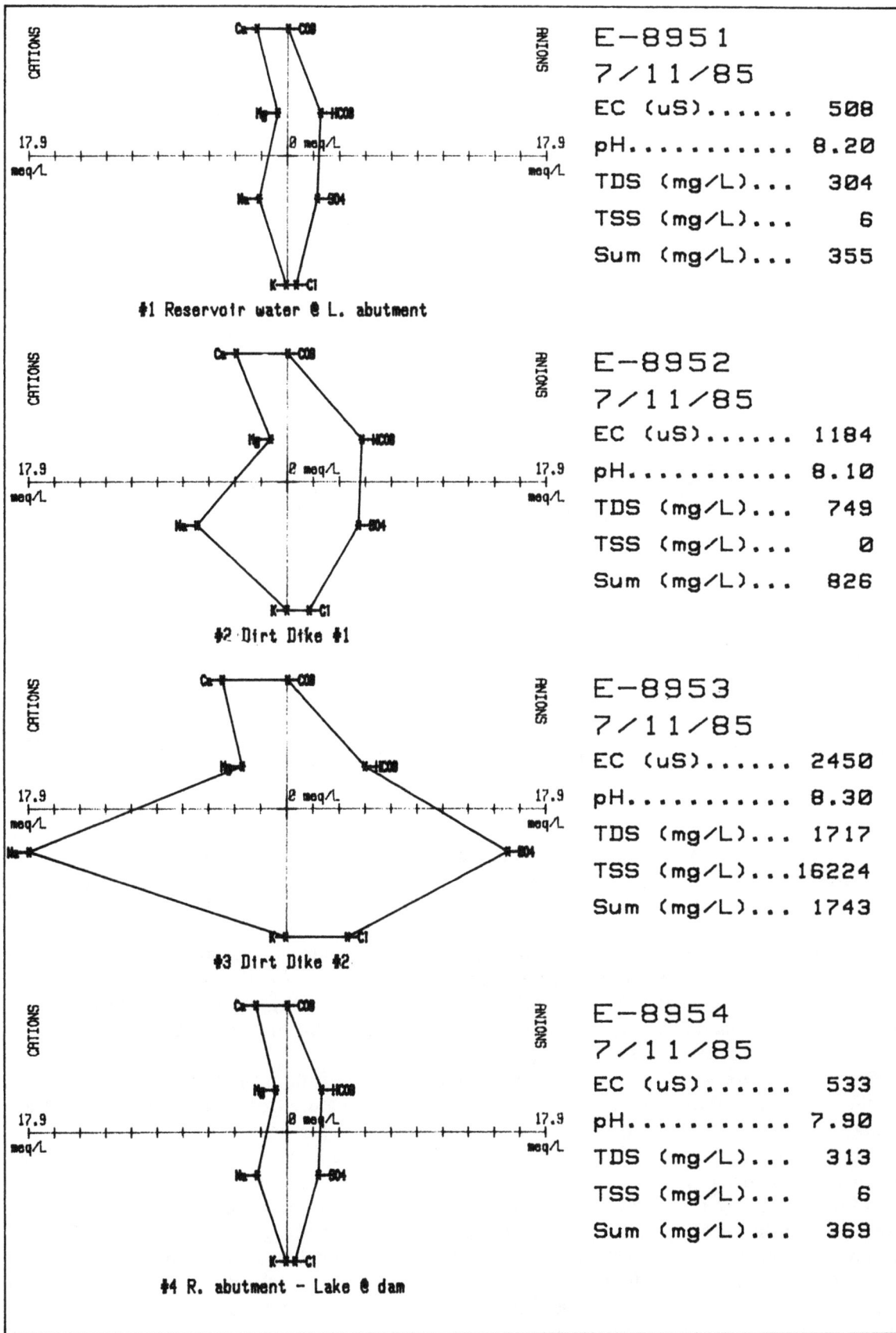

E-8951
7/11/85
EC (uS)......	508
pH..........	8.20
TDS (mg/L)...	304
TSS (mg/L)...	6
Sum (mg/L)...	355

#1 Reservoir water @ L. abutment

E-8952
7/11/85
EC (uS)......	1184
pH..........	8.10
TDS (mg/L)...	749
TSS (mg/L)...	0
Sum (mg/L)...	826

#2 Dirt Dike #1

E-8953
7/11/85
EC (uS)......	2450
pH..........	8.30
TDS (mg/L)...	1717
TSS (mg/L)...	16224
Sum (mg/L)...	1743

#3 Dirt Dike #2

E-8954
7/11/85
EC (uS)......	533
pH..........	7.90
TDS (mg/L)...	313
TSS (mg/L)...	6
Sum (mg/L)...	369

#4 R. abutment - Lake @ dam

Figure 7-10.—Typical stiff diagram for water samples.

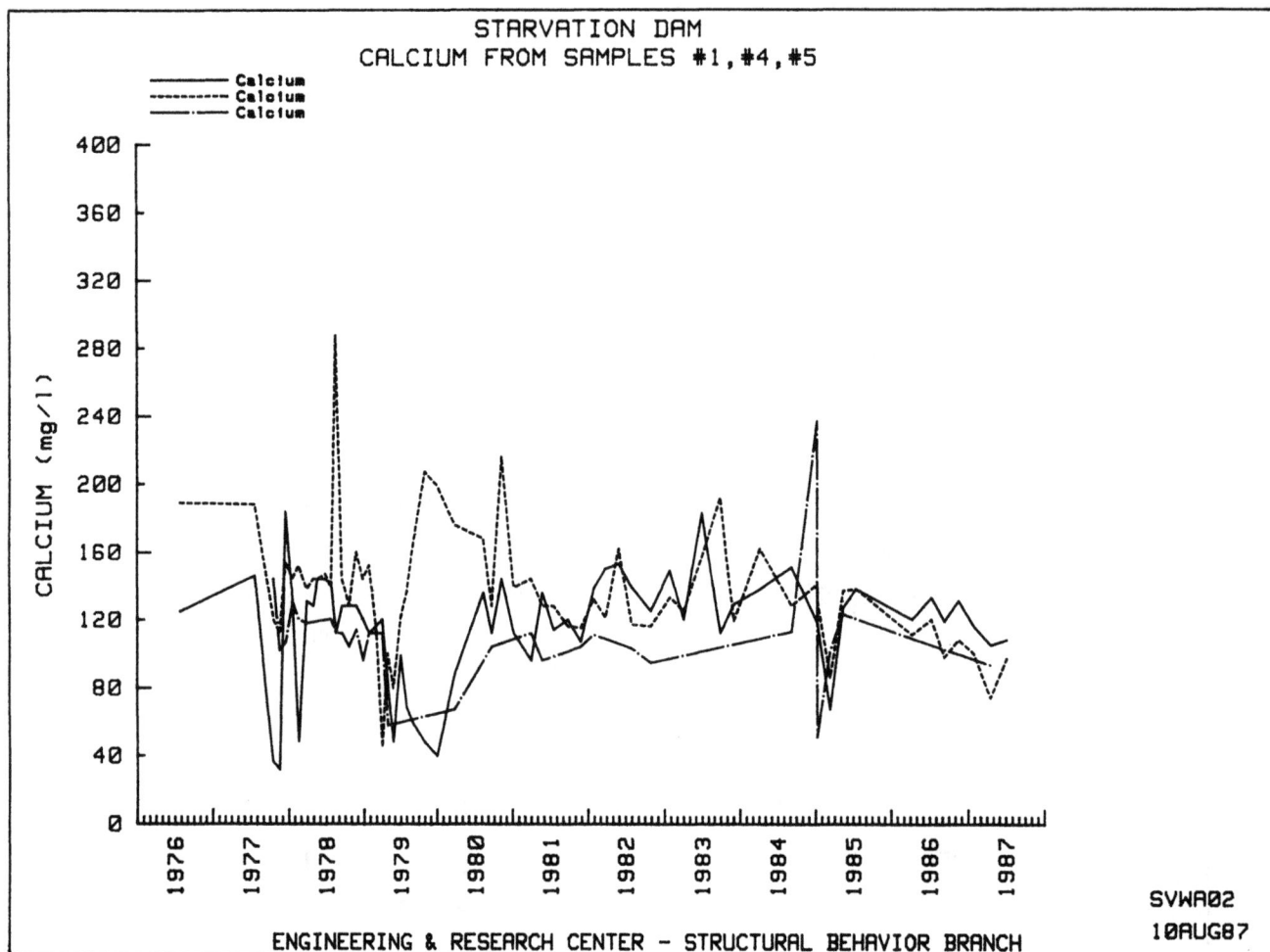

Figure 7-11.—Data plots of water quality analyses.

The automation process begins with an economic analysis to determine if automation can be justified on a labor-saved basis. Selection of the instruments or parameters to be read and the accuracy required from each sensor are included in this analysis. Several other factors such as operating environment, power requirements, interfacing, and where the data obtained should be used must also be addressed. In all cases, the parameters to be measured must be transferred into an electrical analog or digital signal so that they can be scaled into engineering units, transmitted, stored, and analyzed.

The following subsections described several methods of electronically sensing concrete dam structural behavior instrumentation. The electrical signals described are required for remote sensing data acquisition and transmission equipment discussed in section 7.6.

   a. *Pressure Measurement Devices.*—Open piezometers and observation wells in a dam or foundation are good applications for vibrating-wire and resistance strain gauge pressure transducers. These sensors are lowered below the water surface to a predetermined elevation, and a change in the water surface elevation is reflected as a change in the hydraulic head on the transducer's diaphragm with a corresponding change in the electrical output signal. A miniature piezo-resistive pressure transducer has been developed to fit in a ½-inch (13-mm) diameter piezometer standpipe. Like many low-head pressure transducers, this transducer has a vent tube in the power/signal cable to compensate for atmospheric pressure.

Vibrating-wire pressure transducers are of the plucked or auto-resonating type. The plucked type transducer requires a swept frequency to excite the tensioned wire attached between the diaphragm and the transducer body. A pickup coil then senses the natural frequency of the wire at a particular pressure-induced tension. The auto-resonating type sensor has the excitation circuitry housed in the transducer body, and requires direct-current power. The output is a sinusoidal signal that has a frequency related to the sensed pressure.

140

For the pneumatic piezometer, transforming the reading into an electrical signal requires process control techniques. Normally, there are several instruments to be read at the instrument house location. The reading steps require pressurizing the supply line of the piezometer to be read, detecting when the supply gas and the pressure on the piezometer tip are in balance, and then sensing the pressure by an electrical transducer. Desired steady-state flow conditions could be determined by waiting a set time period, such as 5 minutes, or by sensing when pressures stop varying (oscillating) to a significant degree. The supply gas is then depressurized and the next piezometer selected. In some applications, the instruments are kept pressurized at all times, which requires larger amounts of supply gas but allows for faster sampling rates than pressurizing each sensor separately.

b. *Seepage Measuring Devices.*—Seepage is an important parameter when analyzing the condition of foundations and abutments. Seepage water is generally channeled through weirs or, possibly, flumes. To measure flow through a weir, a transducer is submerged upstream and beyond the weir's area of influence. Because a transducer would restrict flow through a measuring flume, the transducer is submerged in a stilling well connected to the flume by a pipe. Both types of measuring devices require two equations to compute the flow. Initially, the height of water above the bottom of the weir notch or the floor of the flume is computed from the relationship between the water depth and the voltage output of the transducer. The flow is then computed using the water height versus the flow calibration for the particular device.

c. *Lake and Tailwater Elevation Devices.*—The difference in head between the lake and tailwater provides the potential for seepage at damsites. If there is a previously installed float, tape, and counterweight-driven chart recorder at a dam, the most direct way to automate the readings is by the transfer of the rotation of the drive pulley to a rotary encoder through sprockets and a chain. The digital output of the encoder is then transmitted as a scaled value. The advantage of this system is that the scale is controlled by the size of the sprockets and the zero offset can be set through switches on the instrument. Only the offset switches need to be changed to keep the encoder and chart recorder readings equal. The mercury manometer system used at some locations may also be automated.

In the absence of chart recording equipment, pressure transducers may be used as previously discussed. However, transducers are more difficult to maintain in calibration than mechanical systems that operate over the large ranges required for lake and tailwater readings. Special methods have been developed to handle the drift and hysteresis characteristics of the transducer.

The electronics required to measure inflow and outflow from the reservoir parallels the techniques used to measure seepage and lake elevation. Extra protection such as a wellhouse in the streambank is required to protect the measuring system from damage due to ice and/or debris. Existing structures such as USGS stream gauging stations already exist on many tributaries.

In general, nearly every type of instrument in use today could be automated to at least some extent. Obviously, economic considerations usually control such efforts.

7.6 Remote Sensing Devices and Methods.—The advent of more economical, microprocessor-controlled, data acquisition equipment suitable for the field environment has made automatic data acquisition more attractive for monitoring dams. The development of a broad line of geotechnical sensors, some of which have been previously discussed, provide electrical signals to be processed by this equipment. The signal processing varies from hand-held units that sense signals from only a few transducers to distributed systems that handle hundreds of sensors and provide satellite and telephone transmission capabilities.

a. *Automatic Data Acquisition.*—The proposed use and the environment will usually dictate the type of data acquisition equipment required on a project. Small projects where only a few transducers are read on a daily or weekly basis might only require a hand-held unit with sensor excitation and digitizing capabilities, and a printer for recording the readings. This type of unit can be powered by rechargeable batteries, and has a sealed housing and keyboard for protection against moisture and dust.

Intermediate systems are usually configured to read 5 to 25 transducers in a local area of interest, and can contain a printer and tape cassette recorder for data storage. Data on the tape are then transferred to a computer for processing. Intermediate data acquisition systems are useful when data must be taken at rates greater than that possible with available personnel, such as a well drawdown test.

When automating large dams and appurtenant structures, multiunit monitors integrated with a central data collection station are found to be the most advantageous. Because the sensors to be monitored at these dams may have to cover a wide area, the clustering of 20 to 50 transducers at one remote monitor that, in turn, communicates the data to a central processor is more economical than connecting all the sensors to a central scanner. The operation of a multiunit system is not dependent on one reading station whose failure would cause total system failure.

Transmission from individual monitors to a central processor is usually performed by a two-wire serial data link cable or radio frequency transmission. The wire link is preferred when the distance to the central station is only about 200 to 300 feet (61 to 91 m). If the distance is greater or the terrain difficult, the radio link becomes more attractive.

There are several methods of protecting remote data acquisition systems from the environment found at a dam. Where instrument houses are provided, the unit should be placed in an enclosure designed to reduce or eliminate excessive moisture. If there are no shelters present, below ground enclosures made of plastic pipe or other noncorrosive material can be provided. It is important that these enclosures be tested prior to installation.

The remote monitors must also be resistant to the electrical environment found at a dam. While the units are required to detect changes in sensor output of a few thousandths of a volt, they must be protected from lightning strikes and other electrical surges that can produce thousands of volts. Some points to be considered when reviewing a system design include isolation from earth ground, use of transformer couplings, varistors, gas discharge tubes, and battery power for protection.

Remote monitors must be able to handle a multitude of sensor types, including common voltage ranges from millivolts to several volts full scale, resistance measurements for potentiometric devices, vibrating-wire excitation and sensing, milliampere detection, and encryption of digital data from rotary encoders. Excitation power strobing should be provided for reading transducers. To ensure correct readings, some sensors require a warmup delay period after excitation power is applied.

The central processor performs several communication functions that include acting as the system communications controller for setting scan intervals at remote monitors, setting system polling intervals, receiving out-of-limit warnings from the remotes, handling remote terminal requests from telephone connections, providing microwave and satellite transmission of blocks of data, and placing warning messages to a preprogrammed telephone number if preset parameters are exceeded. The central processor must also handle contention problems between different communication elements that require protocol algorithms and setting up priorities.

The data processing performed at the central processor includes collection of data from remote units, linear and nonlinear scaling into engineering units, setting warning limits to be used for reporting anomalies, performing data smoothing, and displaying histograms and other statistical data processing. Storage for long-term data retrieval is required to produce time-history displays.

b. Satellite Transmission.—An option for transmitting data to the CYBER computer data files at the Bureau's Engineering and Research Center in Denver is to utilize the GOES-West satellite. The Division of Atmospheric Resources Research currently has transmission time intervals for the satellite that are not presently used and are available to other potential users. Satellite transmission generally operates as follows:

1. Data at a remote location are collected at a transmitter location and transmitted to the satellite using a crossed YAGI antenna. The data are transmitted during set 1-minute transmission periods. These transmission periods can be hourly, or less frequently, depending upon the amount of data to be transmitted. Transmissions begin with the "Transmitter Identification," followed by the data, and then the "Sign Off" signal is given.
2. The satellite relays the data to the satellite down-link at the Engineering and Research Center. The satellite operates at 110-baud which is, unfortunately, quite slow.
3. Data are received at the Division of Atmospheric Resources Research satellite down-link and then transferred to the CYBER computer data files.

Reliability of data transmission by the satellite is considered to be very high.

142

BIBLIOGRAPHY

[1] Jansen, R.B., *Dams and Public Safety*, U.S. Department of the Interior, Bureau of Reclamation, Revised Reprint, Denver, CO, 1983.

[2] *Safety Evaluation of Existing Dams*, U.S. Department of the Interior, Bureau of Reclamation, Revised Reprint, Denver, CO, 1983.

[3] Bartholomew, C.L., B.C. Murray, and D.L. Goins, *Embankment Dam Instrumentation Manual*, U.S. Department of the Interior, Bureau of Reclamation, Denver, CO, 1987.

[4] *Water Measurement Manual*, U.S. Department of the Interior, Bureau of Reclamation, Revised Reprint, Denver, CO, 1984.

[5] Viksne, A., "Bureau of Reclamation Strong Motion Instrumentation Program," *Lifeline Earthquake Engineering*, American Society of Mechanical Engineers, pp. 265-275, 1979.

[6] Hudson, D.E., "Strong Motion Accelerograph Systems—Problems and Prospects," *Proceedings of the Eighth World Conference on Earthquake Engineering*, vol. 2, 1984.

[7] *Uniform Building Code*, "Seismic Zone Map of the United States," 1982.

[8] *Earth Manual*, vol. 2, "Test Designations," Bureau of Reclamation, Denver, CO, (Currently, 1987, in preparation).

[9] *Concrete Manual*, vol. 2, "Test Designations," Bureau of Reclamation, Denver, CO, (Currently, 1987, in preparation).

[10] Raphael, J.M., and R.W. Carlson, *Measurement of Structural Action in Dams*, 2nd ed., Revised April 1956, James J. Gillick and Co., Berkely, CA, 1956.

APPENDIX A
SELECTED BIBLIOGRAPHY

Bartholomew, C.L., L. Johnson, J. Truby, and P. Hagan, *Dam Safety: An Owners Guidance Manual*, Federal Emergency Management Agency, Washington, DC, January 1987.

Bjerrum, L.A., R.B. Casagrande, R.B. Peck, and A.W. Skempton, *From Theory to Practice in Soil Mechanics, Selections From the Writings of Karl Terzaghi*, John Wiley and Sons, Inc., New York, NY, 1960.

Birman, J.H., A.B. Esmilla, and J.B. Indreland, "Thermal Monitoring of Leakage Through Dams," *Geol. Soc. America Bull. 82*, pp. 2261-2284, 1971.

Blake, W., and F. Leighton, "Recent Developments and Applications of the Microseismic Method in Deep Mines," In: *Rock Mechanics: Theory and Practice,* Proceedings, 11th Symposium of Rock Mechanics, American Institute of Mining, Metallurgy and Petroleum Engineers, pp. 429-443, New York, NY, 1970.

Bromwell, L.G., C.R. Ryan, and W.E. Toth, "Recording Inclinometer for Measuring Soil Movement," *Proceeings, 4th Pan American Conference of Soil Mechanics and Foundation Engineering,* vol. 2, pp. 33-343, San Juan, PR, 1971.

Cartwright, K., "Thermal Prospecting for Groundwater," *Water Resources Res.* 4(2), pp. 395-401, 1968a.

Cartwright, K., "The Effect of Shallow Groundwater Flow Systems on Rock and Soil Temperatures," Dissertation, Univ. of Illinois, 1973.

Cording, E.J., A.J. Hendron, Jr., W.H. Hansmire, J.W. Mahar, H.H. MacPherson, R.A. Jones, and T.D. O'Rourk, "Methods for Geotechnical Observations and Instrumentation in Tunneling," Department of Civil Engineering, Univ. of Illinois, Urbana-Champaign, *National Science Foundation,* vol. 2, pp. 293-566, 1975.

Cornforth, D.H., "Performance Characteristics of the Slope Indicator Series 200–B Inclinometer, Field Instrumentation in Geotechnical Engineering," British Geotechnical Society, pp. 126–135, John Wiley and Sons, Inc., New York, NY, 1973.

DiBiago, E., Discussion, *Proceedings, Symposium on Field Instrumentation in Geotechnical Engineering,* British Geotechnical Society, John Wiley and Sons, Inc., pp. 565-566, New York, NY, 1974.

DiBiagio, E., "Field Instrumentation—A Geotechnical Tool," *Norwegian Geotechnical Institute Publication No. 115*, pp. 29-40, 1977.

Dunnicliff, C.J., "Equipment for Field Deformation Measurements," *Proceedings, 4th Pan American Conference on Soil Mechanics and Foundation Engineering,* San Juan, American Society of Civil Engineers, vol. 2, pp. 319-332, New York, NY, 1971.

Dunnicliff, C.J., "Equipment and Methods for Measurement of Displacements and Settlement," Lecture Notes, Met. Section ASCE, Seminar on Field Observations in Foundation Design and Construction, New York, NY, April 1970.

Dunnicliff, C.J., "Schematic Arrangements of Various Types of Soil Mechanics and Rock Mechanics Measuring Instruments," *Highway Focus,* U.S. Department of Transportation, vol. 4, No. 2, June 1972.

Dunnicliff, C.J., *Geotechnical Instrumentation for Monitoring Field Performance,* National Cooperative Highway Research Program, 46 pp., Synthesis of Highway Practice 89, Transportation Research Board, 2101 Constitution Ave. Washington, DC, 20418, 1982.

Dunnicliff, C.J., and J.B. Sellers, "Notes for Training Course on Geotechnical Instrumentation," Implementation Division, Offices of Research and Development, Federal Highway Administraiton, U.S. Department of Transportation, Washington, DC., 1980.

Dutro, H.B., and R.O. Dickinson, "Slope Instrumentation Using Multi-Position Borehole Extensometers," Transportation Research Board, *Transportation Research Record 482,* pp. 9-17, 1974.

Dutta, P.K., R.W. Hatfield, and P.W. Runstadler, Jr., *Calibration Characteristics of Irad Gage Vibrating Wire Stressmeter at Normal and High Temperatures,* Tech. Rep. 80-2, Irad Gage, Lebanon, NH, 1981.

Evaluation of Commercial Soil Pressure Cells, Research Report No. M&R 636342, Sate of California, Materials and Research Department, 1968.

Evaluation of the Portable Borehole Deflectometer, Division of Highways, State of California, Materials and Research Department, Research Report No. M&R 632722-1, April 1968.

Field Instrumentation in Geotechnical Engineering, British Geotechnical Society, John Wiley & Sons, Inc., New York, NY 1974.

Flynn, T.J., S.E. Silliman, and E.S. Simpson, "Water Temperature as a Groundwater Tracer in Fractured Rock," AGU Front Range Conference, April 1985.

"General Considerations on the Measurements for Structural Behavior of Embankment Dams," Report of Measurements Committee, *USCOLD Newsletter,* Issue No. 43, pp. 7-13, available from USCOLD, New York, NY, March 1974.

Gibson, R.E., "An Analysis of System Flexibility and Its Effect on Time-Lag in Pore-Water Pressure Measurements," *Geotechnique,* vol. 13, No. 1, pp. 1-11, 1963.

Gould, J.P., and C.J. Dunnicliff, "Accuracy of Field Deformation Measurements," *Proceedings, 4th Pan American Conference on Soil Mechanics and Foundation Engineering,* vol. II, pp. 313-366, San Juan, PR, 1971.

Green, G.E., "Report on Tests on the Performance of the Wilson Slope Indicator and the Soil Instruments Inclinometer," Unpublished Report, Department of Civil Engineering, Imperial College, London, 1969.

Hansmire, W.H., "Suggested Methods for Monitoring Rock Movements Using Borehole Extensometers," Int. Soc. Rock Mech. Comm. on Standardization of Lab. and Field Tests, 1977.

Hartman, B.E., *Rock Mechanics Instrumentation for Tunnel Construction,* Terrametrics, Inc., Golden, CO, 1967.

Hawkes, I., and J.B. Sellers, *Recent Developments in Multipoint Borehole Extensometers,* 1st Geotechnical Exposition, Chicago, IL, 1979.

"Instrumentation for Measurement of Structural Behavior of Concrete Gravity Structures," *Engineer Manual* EM-1110-2-4300, Department of the Army, Office of the Chief of Engineers, Washington, DC, August, 1970.

"Instrumentation of Earth and Rockfill Dams, Part I, Groundwater and Pore Pressure Observations (1971), Part II, Earth-Movement and Pressure Measuring Devices," *Engineer Manual* 1110-2-1908, Department of the Army, Office of the Chief of Engineers, 1976.

Jacobsen, L.S., and R.S. Ayre, *Engineering Vibrations,* McGraw-Hill, New York, NY, 1958.

Jennings, P.C., and D.V. Helmberger, "Strong Motion Seismology," *Proceedings of the Second International Conference on Microzonation,* vol. 1, pp. 27–53, San Francisco, CA, Nov. 26-Dec. 1, 1978.

Kobold, F., "Measurement of Displacement and Deformation by Geodetic Methods," *Proceedings,* ASCE, vol. 87, No. SU2, Paper 2873, 1961.

MacDuff, J.N., and J.R. Curreri, *Vibration Control,* McGraw-Hill, New York, NY, 1958.

Matthiesen, R.B., (untitled), *Proceedings, Fifth National Meeting of the Universities Council for Earthquake Engineering Research,* Report No. UCEER-5, California Institute of Technology, Mail Code 104-44, pp. 8-10, Pasadena, CA, June 1978.

McVey, J.R., and T.O. Meyer, "An Automatic Data Acquisition System for Underground Measurements," *Report of Investigations 7734,* Spokane Mining Research Center, U.S. Bureau of Mines, 1973.

Measurements of Negative Pore Pressures of Unsaturated Soil Shear and Pore Pressure Research, Bureau of Reclamation Report No. EM-738, Denver, CO, October 11, 1966.

Milner, R.M., "Accuracy of Measurement with Steel Tapes," Current Paper 51/69, Building Research Station, Watford, England, December 1969.

Muller, G., and L. Muller, *Monitoring of Dams with Measurement Instruments,* Trans. 10th Intl. Cong. on Large Dams, vol. III, pp. 1033-1046, Montreal, 1970.

Newmark, N.M., "Design of Structures for Dynamic Loads Including the Effects of Vibration and Ground Shock," Department of Civil Engineering, University of Illinois, Urbana, IL, 1963.

Ogilvy, A.A., M.A. Ayed, and V.A. Bogoslovsley, "Geophysical Studies of Water Leakage from Reservoirs," *Geophysical Prospecting,* vol. 17, pp. 36-62, 1969.

Peck, R.B., "Advantages and Limitations of the Observational Method in Applied Soil Mechanics," *Geotechnique,* vol. 19, No. 2, pp. 171-187, 1969.

Peck, R.B., "Observation and Instrumentation: Some Elementary Considerations," *Highway Focus,* vol. 4, No. 2, pp. 1-5, 1972.

Peck, R.B., "Influence of Nontechnical Factors on the Quality of Embankment Dams," *Embankment Dam Engineering, Casagrande Volume,* John Wiley and Sons, Inc., pp. 201-208, New York, NY, 1973.

Penman, A.D.M., "A Study of the Response Time of Various Types of Piezometers," *Proceedings, Conference on Pore Pressure and Suction in Soils,* pp. 53-58, Butterworths and Co., London, 1960.

Penman, A.D.M., "Instrumentation for Earth and Rockfill Dams," Building Research Station, Current Paper 35/69, Watford, England, September 1969.

Procedure for Installation of Positive Operating Piezometer, Inspector's Booklet, Piezometer Research & Development Corp., Stamford, CT, 1968.

"Recommended Guidelines for Safety Inspection of Dams, National Program of Inspection of Dams, Appendix D, Department of the Army, Office of the Chief of Engineers, Washington, DC, May 1975.

Safety of Existing Dams, Evaluation and Improvement, National Research Council, Committee on Safety of Existing Dams, National Academy Press, 1983.

Sherard, J.L., "Piezometers in Earth Dam Impervious Sections," *Recent Developments in Geotechnical Engineering for Hydro Projects,* Fred H. Kulhawy, editor, ASCE, New York, NY, May 1981.

Standard Methods for the Examination of Water and Wastewater, American Public Health Association, 15th ed., 1981.

Statistical Compilation of Storage Dams, Dikes, and Reservoirs on Bureau of Reclamation Projects, (also available for diversion dams, major pumping plants, and carriage facilities-tunnels, canals, and pipelines), published annually by the United States Department of the Interior, Bureau of Reclamation, Denver, CO, 1978.

Stiff, H.A., Jr., "The Interpretation of Chemical Water Analysis by Means of Diagrams," *Journal of Petroleum Technology,* vol. 3, No. 10, 1951.

"Suggested Methods for Monitoring Rock Movements Using Inclinometers and Tiltmeters," Commission on Standardization of Laboratory and Field Tests, International Society of Rock Mechanics, *Rock Mechanics,* vol. 10/1-2, pp. 81-106, 1977.

Terzaghi, K., "Mechanism of Landslides," *Application of Geology to Engineering Practice,* S. Paige, editor, Geological Society of America, Berkey Edition, vol. 1950, pp. 83-123.

Terzaghi, K., and R.B. Peck, *Soil Mechanics in Engineering Practice,* 2d ed., ch. 12, John Wiley & Sons, Inc., New York, NY, 1967.

Vaughn, P.R., "A Note on Sealing Piezometers in Boreholes," *Geotechnique,* vol. 19, No. 3, pp. 405-413, 1969.

Vaughn, P.R., "The Measurement of Pore Pressures with Piezometers," *Symposium on Field Instrumentation in Geotechnical Engineering*, John Wiley & Sons, Inc., pp. 422-441, New York, NY, 1974.

Ward, W.H., and J.E. Cheney, "Oscillator Measuring Equipment for Vibrating-Wire Gages," *Journal Scientific Instruments,* vol. 37, pp. 88-92, March 1960.

Wilkes, P.F., "The Installation of Piezometers in Small Diameter Boreholes," *Geotechnique,* vol. 20, No. 3, pp. 330-333, 1970.

Wilson, S.D., "The Use of Slope Measuring Devices to Determine Movements in Earth Masses," In: *Field Testing of Soils,* American Society for Testing and Materials, Special Technical Publications 322, pp. 187-197, 1962.

Wilson, S.D., and P.E. Mikkelsen, *Foundation Instrumentation—Inclinometers Reference Manual,* FHWA, Report No. FHWA TS-77-219, 1977.

Wilson, S.D., and P.E. Mikkelsen, "Field Instrumentation, Landslide Analysis and Control," Transportation Research Board, Special Report 176, pp. 112-138, 1978.

APPENDIX B

Summary of Instrumentation in Bureau of Reclamation Concrete Dams

Name of Dam	Year Completed	Height in feet (meters)	Crest Length in feet (meters)	Type of Instrumentation
Altus	1945	111 (33.8)	1,112 (338.9)	Uplift pressures Piezometers Seepage/drainage Collimation
American Falls	1978	103.5 (31.5)	5,277 (1608.4)	Collimation Seepage/drainage Uplift pressures Embankment measuring points
Angostura	1949	193 (58.8)	2,030 (618.7)	Seepage/drainage Uplift pressures
Bartlett	1939	287 (87.5)	800 (243.8)	Collimation Levels Crack monitors
Buffalo Bill	1910	325 (99.1)	200 (61)	Piezometers
Black Canyon	1924	183 (55.8)	1,039 (316.7)	Seepage/drainage Uplift pressures
Brantley	UC[1]	143.5 (43.7)	20,900 (6370.3)	Levels Whittemore points MPBX[2] Uplift pressures Thermometers Joint meters EDM[3] Seepage/drainage
Canyon Ferry	1954	225 (68.6)	1,000 (304.8)	Uplift pressures Joint meters Strain meters Penstock meters Reinforcement meters No-stress strain meters Drain flow
Crystal	1977	323 (98.5)	635 (193.5)	Collimation Foundation deformation Plumblines Thermometers Tape gauges Slide movements Drain flows Whittemore gauges Formed drains
East Canyon	1966	260 (79.2)	436 (132.9)	Collimation Foundation deformation Meter circuit resistance Triangulation Seepage/drainage

Name of Dam	Year Completed	Height in feet (meters)	Crest Length in feet (meters)	Type of Instrumentation
Elephant Butte	1916	301 (91.7)	1,674 (510.2)	Seepage/drainage Piezometers Levels Collimation Crack monitors EDM[3]
Flaming Gorge	1964	502 (153.0)	1,285 (391.7)	Plumblines Seepage/drainage EDM[3] Uplift pressures Thermometers Joint meters
Folsom	1956	340 (103.6)	1,400 (426.7)	Seepage/drainage Uplift pressures Joint meters
Friant	1942	319 (97.2)	3,488 (1063.1)	Thermometers Contraction joints Seepage/drainage Plumblines Uplift pressures
Grand Coulee	1942	550 (167.6)	5,223 (1592.0)	Seepage/drainage Uplift pressure
Grand Coulee Forebay	1974	200 (61.0)	1,170 (356.6)	Seepage/drainage Uplift pressures Whittemore gauges Collimation Embedded instruments Deflectometers Foundation deformation Plumblines Thermometers
Gibson	1929	199 (60.7)	960 (292.6)	Collimation Uplift pressures
Glen Canyon	1964	710 (216.4)	1,560 (475.5)	Seepage/drainage Uplift pressures Plumblines Embedded instruments Abutment deformation EDM[3]
Hungry Horse	1953	564 (171.9)	2,115 (644.7)	Drain flows Uplift pressures Plumblines
Horse Mesa	1927	305 (93.0)	660 (201.2)	Collimation Levels Crack monitors
Hoover	1936	726 (221.3)	1,244 (379.2)	Drain flows Uplift pressure

Name of Dam	Year Completed	Height in feet (meters)		Crest Length in feet (meters)		Type of Instrumentation
Keswick	1950	157	(47.9)	1,046	(318.8)	Seepage/drainage Uplift pressure
Kortes	1951	244	(74.4)	440	(134.1)	Seepage/drainage Uplift pressure
Lake Tahoe	1913	18	(5.5)	109	(33.2)	Piezometers
Monticello	1957	304	(92.7)	1,023	(311.8)	Plumblines EDM[3] Strain meters Triangulation
Marshall Ford	1942	278	(84.7)	5,093	(1552.3)	Uplift pressures Drain flows Embankment measuring points Benchmark settlement
Mormon Flats	1926	224	(68.3)	380	(115.8)	Collimation Levels Whittemore points
Mountain Park	1975	133	(40.5)	320	(97.5)	Collimation Thermometers
Morrow Point	1968	468	(142.6)	724	(220.7)	Collimation Foundation deformation Drain flows Plumblines Penstock meters EDM[3] Thermometers Strain meters Stress meters Joint meters No-stress meters Landslide movement
Nambe Falls	1976	150	(45.7)	320	(97.5)	Collimation Drain flows Uplift pressures Joint meters Flat jack pressures Observation wells Thermometers Strain meters Piezometers Settlements
Nimbus	1955	87	(26.5)	1,093	(333.1)	Collimation Seepage/drainage Piezometers Levels
Olympus	1949	70	(21.3)	1,951	(594.7)	Seepage/drainage Uplift pressures

Name of Dam	Year Completed	Height in feet (meters)	Crest Length in feet (meters)	Type of Instrumentation
Owyhee	1932	417 (127.1)	833 (253.9)	Crack monitoring Uplift pressures Seepage/drainage EDM[3] Levels
Parker	1938	320 (97.5)	856 (260.9)	Collimation Tape gauges Levels MPBX[2]
Pathfinder	1909	214 (65.2)	432 (131.7)	Seepage/drainage Levels Collimation
Pueblo	1975	191 (58.2)	10,200 (3110.0)	Plumblines Collimation Seepage/drainage Uplift pressures Buttress movements Foundation deformation Piezometers Observation wells Thermometers Whittemore gauges Deflectometers EDM[3] Embankment points
Red Bluff	1964	52 (15.8)	5,985 (1824.2)	Pore pressures
Seminoe	1939	295 (89.9)	530 (161.5)	Uplift pressures Settlements Drain flows EDM[3]
Stony Gorge	1928	139 (42.4)	868 (264.6)	Collimation Whittemore gauges Observation wells Thermometers Levels
Shasta	1945	602 (183.5)	3,460 (1054.6)	Seepage/drainage Plumblines Uplift pressures
Stewart Mountain	1930	207 (63.1)	1,260 (384.0)	Collimation Whittemore gauges Levels Powerplant movements EDM[3]
Theodore Roosevelt	1911	280 (85.3)	723 (220.4)	Collimation Levels MPBX[2] Piezometers

Name of Dam	Year Completed	Height in feet (meters)		Crest Length in feet (meters)		Type of Instrumentation
Upper Stillwater	UC[1]	286	(87.2)	2675	(815.3)	Observation wells Thermometers Piezometers Stress meters Inclinometers
Warm Springs	1919	106	(32.3)	469	(143.0)	Collimation
Yellowtail	1966	525	(160.0)	1,480	(451.1)	Plumblines Collimation Seepage/drainage Uplift pressures Spring flows Strain meters Foundation deformation Observation wells Thermometers

[1]UC indicates under construction.
[2]MPBX indicates multiple-point borehole extensometer.
[3]EDM indicates electronic distance measurement.

☆ U.S. GOVERNMENT PRINTING OFFICE:1988-573-907/85030

www.ingramcontent.com/pod-product-compliance
Lightning Source LLC
Chambersburg PA
CBHW061326190326
41458CB00011B/3904

9 781780 393612